Principles and Applications of Up-converting Phosphor Technology

Ruifu Yang
Editor

Principles and Applications of Up-converting Phosphor Technology

Science Press
Beijing

Springer

Editor
Ruifu Yang
State Key Laboratory of Pathogen
and Biosecurity
Beijing Institute of Microbiology
and Epidemiology
Beijing, China

ISBN 978-981-32-9281-9 ISBN 978-981-32-9279-6 (eBook)
https://doi.org/10.1007/978-981-32-9279-6

Jointly published with Science Press
The print edition is not for sale in China. Customers from China please order the print book from: Science Press.

This Springer imprint is published by the registered company Springer Nature Singapore Pte Ltd.
The registered company address is: 152 Beach Road, #21-01/04 Gateway East, Singapore 189721, Singapore

Contents

Chapter 1
Upconversion Luminescent Materials: Properties and Luminescence Mechanisms

Dandan Song, Suling Zhao and Zheng Xu

Abstract The upconversion (UC) luminescent materials typically refer to the materials doped with rare earth (RE) ions or transition metal ions as the luminescence centers, among which the lanthanides ions possess superior UC efficiency. UC luminescence (UCL) is a kind of anti-Stokes process, which absorbs two or more photons with a low energy while emits one photon with a high energy. The energy difference between the absorbed and emitted photons are typically much larger than kT (k is the Boltzmann constant and T is the Kelvin temperature). At present, the UCL covers the whole visible spectrum range, which can be applied in many fields including the solid laser, the multicolor display technology, the optical data storage, the biological probe, and the bio-imaging. Especially in the biological field, the UC materials can penetrate a much depth in the body benefiting from its infrared light excitation, and meanwhile, the florescence emission from the organic molecules can be avoided. Therefore, the UC materials yield many unique features in the detection and the identification of the biomolecules. Here, in this chapter, a fundamental description of the UC processes and the properties of the UC materials is provided. Section 1.1 provides a brief introduction to the UCL materials and the main UCL processes. Section 1.2 describes the necessary mechanisms and the processes involved in the UCL. Section 1.3 introduces the methods for improving the luminescent performance of the UCL materials.

Keywords Upconversion luminescence · Excited state absorption · Photon-avalanche · Energy transfer upconversion · Rare earth

---

D. Song · S. Zhao · Z. Xu (✉)
Key Laboratory of Luminescence and Optical Information (Beijing Jiaotong University), Ministry of Education, Beijing 100044, China
e-mail: zhengxu@bjtu.edu.cn

D. Song
e-mail: ddsong@bjtu.edu.cn

S. Zhao
e-mail: slzhao@bjtu.edu.cn

Institute of Optoelectronics Technology, Beijing Jiaotong University, Beijing 100044, China

R. Yang (ed.), *Principles and Applications of Up-converting Phosphor Technology*,
https://doi.org/10.1007/978-981-32-9279-6_1

1.1 Introduction to the Upconversion Luminescence

1.1.1 Upconversion Luminescence

With the materials upon the excitation of certain energy, such as light, heat, electrical stress, and without chemically changed, it will give the excess energy and back to its equilibrium state. If the energy is given as the electromagnetic wave in the visible or near visible range, the phenomenon is called as luminescence (Xu et al. 2004). Luminescent materials include the semiconductor luminescent materials, the materials with discrete luminescent centers, some special structured materials (like quantum well), and so on.

Generally, the material will emit photons with a smaller energy than the absorbed photons, and the difference between the emission spectrum and the absorption spectrum is called as the Stokes shift (as can be seen in Fig. 1.1a). If the material emits photons with higher energy than the absorbed photons, the difference is called as the anti-Stokes shift. The materials with anti-Stokes shift contain UC materials, multiphoton absorption materials, hot band absorption, and triplet-triplet annihilation (Zhu et al. 2017), which possess many unique properties in the application. The different emission processes can be seen from Fig. 1.1b–f. In a Stokes shift process,

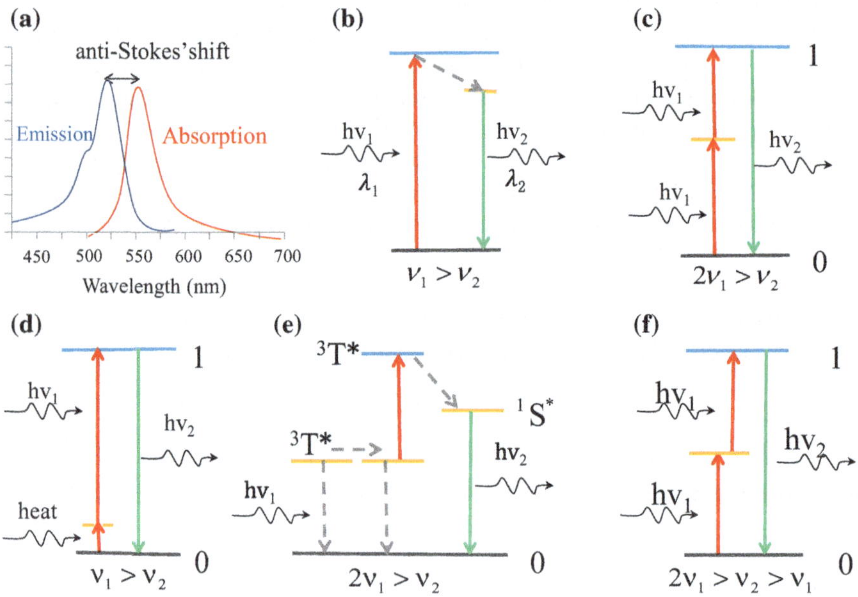

Fig. 1.1 Stokes shift and anti-Stokes shift luminescence processes. **a** Absorption and emission spectra of the Stokes shift materials; **b** Stokes shift luminescence process; **c** two-photon absorption process; **d** hot band absorption process; **e** triplet-triplet annihilation process; **f** UC process (O'Brien 1946)

according to the general luminescent materials, the electron at the ground state captures the energy of the incident photon and jumps to the excited state, and then it back to the ground state by emitting a photon with lower energy and losing the excessive energy. In the two-photon absorption process, the electron absorbs two photons with low energy and jumps to the excited state, followed by emitting one photon with high energy. In the hot band absorption process, the electron absorbs the hot energy and the photon simultaneously, and then emits a photon with high energy. In the triplet-triplet annihilation process which mainly occurs in the organic materials, two triplet excitons with low energy generate one singlet exciton with high energy.

In the UC process, the electrons capture two or more photons by absorption or energy transfer, and then gives photons with high energy when electrons transit to a lower state or the ground states (the energy difference is much larger than kT). The UC process is an important example of the anti-stokes shift, which effectively increases the photon energy by a factor of 2 or more. Hence, the UC process can translate the infrared light to the visible light. Different from other multiple photon absorption process, the UC process can occur at low excitation density.

The discovery and the research of the UCL phenomenon starts from the middle of last century. Before 1940s, it was found that a type of phosphor could emit visible light upon excitation with infrared light (O'Brien 1946), and this phenomenon was defined as the UCL. However, this actually is a kind of infrared stimulated luminescence. In 1959, Bloebergen et al. proposed a design of the infrared quantum counter by translating the infrared photons into visible photons through a sequential absorption process of the ions (Bloembergen 1959), i.e., the sequential absorption by the excited states of the ions, which is the earliest description of achieving UCL via excited state absorption. In the middle of 1960s, Auzel et al. investigated the UCL from the materials co-doped with RE ions (Auzel 1966). They discovered the UCL in the materials with Yb^{3+} ions co-doped with other RE ions such as Er^{3+}, Tm^{3+}, and Pr^{3+}. They attributed the UCL to the addition de Photons par Transferts d'Energie (APTE) i.e., the energy transfer upconversion (ETU) (Cresswell et al. 1978; Zhou et al. 2012; Auzel 1973) by exciting Yb^{3+}. The concept of ETU by other ions to boost UCL increases the UCL efficiency by a large content, and also it makes the realizing of lasers with single frequency pumping to be possible. In 1979, Chivian et al. discovered the photon avalanche (PA) UC process in the infrared quantum counter based on Pr^{3+} ions.

Besides the RE ions, some transition metal ions as the luminescent centers are also found to be able to exhibit UCL. In 1978, Cresswell et al. utilized Re^{4+} ion to replace Yb^{3+} ion in Yb–Tm system in Cs_2NaYCl_6 matrix, which enabled the UC from the infrared light to green light (Cresswell et al. 1978). Auzel et al. observed UC emission from $MgF_2:Ni^{2+}$ (Moncorgé et al. 2006). In addition, Ti^{2+} ($^3d_2$) ion, Cr^{3+} ($^3d_3$) ion, Mo^{3+} ($^4d_3$) ion and Os^{4+} ($^5d_4$) ion are also proved to be able to exhibit UCL (Auzel 2004). However, due to the strong Stokes shift of d ions in the medium crystal field, the UCL from the transition metal ions is not as efficient as that from the RE ions.

The UC materials were mainly utilized in the field of lasers in the initial stage since they were discovered, which could convert the infrared light to the visible laser.

In 1971, Johnson et al. firstly realized the green laser from UCL using flash gun as the pump at 77K utilizing BaY_2F_8: Yb/Ho and BaY_2F_8: Yb/Er UC materials (Xu et al. 2004). In 1987, Antipenko et al. realized the laser by UCL of BaY_2F_8: Er material at room temperature. At present, the lasers in violet, blue, green, and red spectrum are all be realized by using UCL at room temperature, and the highest power reaches 1238 mW and the slope efficiency reaches 46.6% (Fujimoto et al. 2013). The UCL can also be utilized in solar cells, which can convert the infrared part of the solar spectrum to the visible region where the solar cell is able to absorb. Hence, the absorption of the solar spectrum by the solar cells can be improved (Huang et al. 2013). However, due to the relatively low UC efficiency of the UC materials, the improvement in the final efficiency of the solar cells is few. The UCL also possess advantages in other applications, such as 3D displays and anti-forgery technologies.

An extreme important application of the UC materials is in the biological fields. With the developments of the biological technologies, labeling, imaging, and detecting of the biomolecules by means of the luminescence become more and more important. A variety of materials, including fluorescent protein, organic dyes, organometallic compounds, semiconductor quantum dots and so on, can be used as the fluorescence probe. However, the excitation of these materials requires the photons with high energy, which may damage the DNA and cells (Zhou et al. 2012). Moreover, the biomolecules may also be excited simultaneously, which also emit light and disturbs the detection. Therefore, the anti-Stokes shift materials with the ability of converting near-infrared to visible light are of great advantages in this field. The UC nanoparticles with lanthanide metal ions as the luminescent centers are capable of emitting visible light at a low excitation influence, which are successfully applied in biological labeling, imaging, and thereby.

1.1.2 The Main Processes in Upconversion Luminescence

Typically, the luminescence only involves one ground state and one excited state. In UCL, there are many metastable excite states involved. The superposition of the energy from these intermediate states enables the conversion from low energy photons to high energy photons. Ions with f and d levels contain the metastable states, and hence in theoretically, the RE ions with f level (4f, 5f) and the transition metal ions with d level (3d, 4d, 5d) are able to create UCL.

As each type of ions has their certain energy levels, so the UC processes are also different among different ions. There are basically three kinds of UC processes, including the excited state absorption (ESA), the photon-avalanche (PA) and the energy transfer upconversion (ETU, or addition de photon par transferts d'energie, APTE). A schematic illustration of the different UC processes is shown in Fig. 1.2.

The UCL based on ESA process is realized through the sequential absorption at the excited state, as shown in Fig. 1.2a, the electron jumps to the metastable excited state E_1 by absorbing a pump photon and then jumps the excited state E_2 by absorbing another pump photon. The transition from E_2 to the ground state G creates a photon

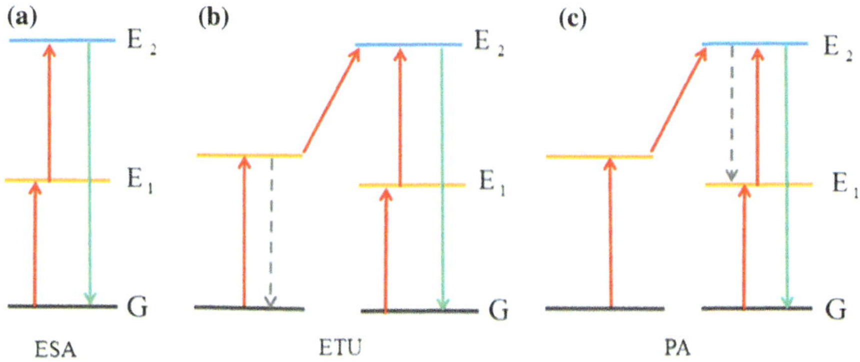

Fig. 1.2 Schematic illustration of the UC processes. **a** Excited state absorption (ESA); **b** energy transfer upconversion (ETU); **c** photon-avalanche (PA). The red, dashed grey and green arrows represent the absorption process, the energy transfer process and the emission process, respectively

with higher energy that equals to the sum of two pump photons. The UCL based on ETU process also involves the metastable excited state absorption, whereas the metastable excited state absorption is assisted by the energy transfer from another RE ion instead of direct absorption of the pump photon in ESA process. As can be seen in Fig. 1.2b, the electron of one RE ion absorbs the pump photon and jumps to the metastable excited state E1, followed by obtaining additional energy from adjacent ion through energy transfer and then jumping to a higher excited state E2, realizing the energy upconversion. The PA process is shown in Fig. 1.2c, the electrons are populated at the metastable excited state by nonresonant absorption and then jump to a higher excited state by the resonant absorption of the pump light.

In these three processes, the UC efficiency of the ESA process is relatively low, and the UC efficiency of the ETU process is the highest. Therefore, the efficient UC materials are based on the ETU process. The performance of some typical UC materials based on different processes are listed in Table 1.1 (Auzel 1973, 1975, 1990; Nakazawa and Shionoya 1970; Auzel and Pecile 1973, 1976; Page et al. 1998; Chamarro and Cases 1988; Wu et al. 1994; Hubert et al. 1986; Salley et al. 2001; Hehlen et al. 2001).

1.1.3 Upconversion Materials

(1) **Catalogues of the UC materials**

In UC materials, ions with the metastable excited states are the luminescent centers, which upconvert the energy. Now the ions with certified UCL are the RE ions and the transition metal ions which contain f or d energy level. Figure 1.3 presents the ions which are capable for UCL in the periodic table of elements, including lanthanides ions (Pr^{3+} ($4f^2$), Nd^{3+} ($4f^3$), Sm^{3+} ($4f^5$), Eu^{3+}

Table 1.1 Performance of the UC materials based on different processes (Auzel 1973, 1975, 1990; Nakazawa and Shionoya 1970; Auzel and Pecile 1973, 1976; Page et al. 1998; Chamarro and Cases 1988; Wu et al. 1994; Hubert et al. 1986; Salley et al. 2001; Hehlen et al. 2001)

Matrix	RE ions	Process	Temp (K)	Effiiency $(cm^2/W)^{n-1}$	References
YF_3	Yb^{3+}–Er^{3+}	APTE (ETU)	300	$\cong 10^{-3}$	Auzel (1973)
SrF_2	Er^{3+}	ESA	300	$\cong 10^{-5}$	Auzel (1973)
YF_3	Yb^{3+}–Tb^{3+}	Cooperative sensitization	300	$\cong 10^{-6}$	Auzel (1973)
$YbPO_4$	Yb^{3+}	Cooperative luminescence	300	$\cong 10^{-8}$	Auzel (1990), Nakazawa and Shionoya (1970)
CaF_2	Eu^{3+}	Two-photon absorption	300	$\simeq 10^{-13}$	Auzel (1973)
YF_3	Yb^{3+}–Er^{3+}	APTE (ETU)	300	2.8×10^{-1}	Auzel and Pecile (1973)
vitroceramics	Yb^{3+}–Er^{3+}	APTE (ETU)	300	2.8×10^{-1}	Auzel (1975)
NaYF4	Yb^{3+}–Tm^{3+}	APTE (ETU)	300	3.4×10^{-2}	Auzel and Pecile (1976)
YF_3	Yb^{3+}–Tm^{3+}	APTE (ETU)	300	4.25×10^{-2}	Auzel and Pecile (1976)
vitroceramics	Yb^{3+}–Tm^{3+}	APTE (ETU)	300	8.5×10^{-2}	Auzel and Pecile (1976)
$NaYF_4$, $Na_2Y_3F_{11}$	Yb^{3+}–Er^{3+}	APTE (ETU)	300	10^{-2} to 2×10^{-4}	Page et al. (1998)
$NaYF_4$	Yb^{3+}–Er^{3+}	APTE (ETU)	300	2.5×10^{-4}	Hehlen et al. (2001)
$NaYF_4$	Yb^{3+}–Tm^{3+}	APTE (ETU)	300	5.5×10^{-2}	Page et al. (1998)
$NaYF_4$	Yb^{3+}–Tm^{3+}	APTE (ETU)	300	3×10^{-7}	Hehlen et al. (2001)
fluorohanfnate glass	Yb^{3+}–Tm^{3+}	APTE (ETU)	300	6.4×10^{-3}	Chamarro and Cases (1988)
fluorohanfnate glass	Yb^{3+}–Ho^{3+}	APTE (ETU)	300	8.4×10^{-4}	Chamarro and Cases (1988)
vitroceramics	Yb^{3+}–Tm^{3+}	APTE (ETU)	300	3.5×10^{-1}	Wu et al. (1994)
vitroceramics	Yb^{3+}–Tm^{3+}	APTE (ETU)	300	3.6×10^{-3}	Wu et al. (1994)
$ThBr_4$	U^{4+}	ESA	300	2×10^{-6}	Hubert et al. (1986)
$SrCl_2$	Yb^{3+}–Yb^{3+}	Cooperative luminescence	100	1.7×10^{-10}	Salley et al. (2001)
$SrCl_2$	Yb^{3+}–Tb^{3+}	Cooperative sensitization	300	8×10^{-8}	Salley et al. (2001)

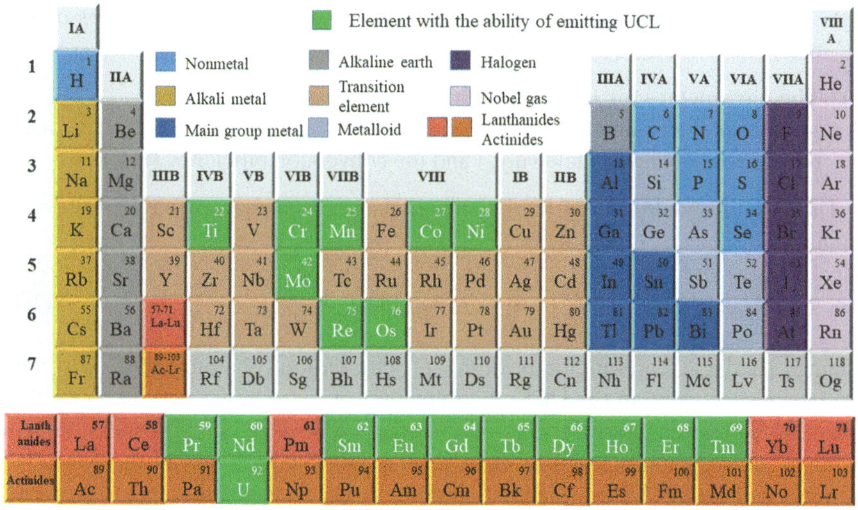

Fig. 1.3 Elements with the ability of emitting UCL in periodic table of elements

($4f^6$), Gd^{3+}($4f^7$), Tb^{3+} ($4f^8$), Dy^{3+} ($4f^9$), Ho^{3+} ($4f^{10}$), Er^{3+} ($4f^{11}$), Tm^{3+} ($4f^{12}$), Tm^{2+} ($4f^{13}$)), actinides ions (U^{4+} (5f2), U^{3+} ($5f^5$)), and transition metal ions (Ti^{2+} ($3d^2$), Cr^{3+} ($3d^3$), Mn^{2+} ($3d^5$), Ni^{2+} ($3d^8$), Cu^{2+} ($3d^9$), Mo^{3+} ($4d^3$), Re^{4+} ($5d^3$), Os^{4+} ($5d^4$)).

RE ions, as the upconversion luminescence centers, include lanthanide and actinide ions. The upconversion luminescence of lanthanide ions is based on the electron transition between 4f energy levels. Due to the inner shell configurations of 4f by 5d and 5p, the influence of the crystal field of different host matrix to the 4f–4f transitions is negligible, which give the stable UCL. When the phonon energy of the host matrix is low, the rate of the multiphonon relaxation will be reduced (Sangeetha and van Veggel 2009). In such case, the lifetime of electrons on metastable excited states will be longer, which will help to realize the high efficient upconversion luminescence. By doping the lanthanide RE ions in different matrix, we can obtain the UCL with different colors (Malinowski et al. 2000). For example, co-doping Tm^{3+} and Yb^{3+} in fluoride glass can create blue emission (Martín et al. 1999; Méndez-Ramos et al. 2001). There are hundreds of UC materials with different ions doped in different matrix.

Different from the RE ions which have narrow emission spectra and certain emission wavelength, the transition metal ions show wide emission spectra. Due to the sensitivity of the d electrons to the chemical surroundings, the emission color of the transition metal ions can be adjusted (Ofelt 1962). The transition metal ions can also be used together with the RE ions to create UCL.

Though the transition metal ions are capable to obtain UCL, the UCL only occurs at low temperature and the non-radiative rate of the transition metal ions is very high, leading to the low UC efficiency. Hence, this chapter mainly focuses on the UC materials consisted of RE ions, especially the lanthanide ions.

(2) **The parameters for evaluating the UC materials**
In general, the parameters for evaluating the UCL performance of the UC materials are the quantum yield (η_{QY}) and the conversion efficiency (η_{CE}).

The quantum yield η_{QY} of the UC materials can be expressed by the following equation

$$\eta_{QY} = n_{em}/n_{abs} \tag{1.1}$$

where n_{em} and n_{abs} represent for the numbers of the emitted photons and the absorbed photons, respectively. In the UCL, the maximum η_{QY} for a two-photon absorption scheme is 50% while for a three-photon absorption scheme is 33%. The measurement of the absolute quantum yield of the UC materials can be found in the published literatures including "Measurement of photoluminescence quantum yields. Review" (*The Journal of Physical Chemistry*, 1971, **75**, 991–1024) (Lee et al. 2004) and "Measuring the absolute quantum efficiency of luminescent materials" (*Journal of luminescence*, 2005, **115**, 77–90) (Rohwer and Martin 2005).

The conversion efficiency η_{CE} of the UC materials can be expressed by the following equation

$$\eta_{CE} = P_{em}/P_{abs} \tag{1.2}$$

where P_{em} and P_{abs} represent for the power of the emitted photons and the absorbed photons, respectively. As the power of the photons is determined by their wavelengths, so the conversion efficiency is not equal to the quantum yield.

Many factors may influence the UC efficiency of the materials, the following are some important ones.

(a) The energy level structure and the doping concentration of the ions.

The energy difference between the higher energy level and the next lower level effects the radiation rate of the higher energy level. The bigger the energy difference is, the lower the non-radiation rate is, which will result a higher radiation rate and then the higher efficient upconversion luminescence, and vice versa. Generally, a higher content of doped ions can cause the concentration quench of the upconversion luminescence due to the cross-relaxation between different ions. In order to avoid this phenomenon, a low content of doped ions and a high excitation can be combined to achieve efficient upconversion luminescence (Kaminskii 2013).

(b) The features of the host matrix.

The effect of the host matrix includes the following aspects. (I) The phonon energy of the host matrix will affect the multiphonon relaxation and the phonon-assisted energy transfer of excited RE ions, which is one of the key factors related with the upconversion efficiency. (II) The crystal symmetry of the host matrix will influence the electric dipole transition probability of lanthanide ions. Therefore, lanthanide ions with different radius are doped to induce the lattice expansion or shrinkage to adjust the crystal symmetry of the host matrix (Han et al. 2014; Zhao et al. 2013). Or co-doping the transition metal ions increases the electron–phonon interaction of lanthanide ions by the d orbit effect of transition metal ions (Han et al. 2014; Tang et al. 2015).

(c) The temperature of the surroundings.

By increasing the temperature, the multiphonon relaxation rate of the higher energy level also increases, whereas the efficiency of upconversion luminescence will decrease. In addition, the energy transfer probability assisted by absorbed phonons will increase and thereby emitted phonons will decrease when the temperature increases.

1.2 The Processes Governing the Upconversion Luminescence

The extensive investigated upconversion materials with excellent performance are RE ions doped upconversion materials. In this section, the upconversion mechanism will be introduced based on the energy level and electron transition of trivalent lanthanide ions.

1.2.1 The Energy Level Structure and the Radiative Transition of Rare Earth Ions (Huang 2002; Yen and Selzer 1981)

The energy level structure and the electron transition between energy levels of RE ions are determined by the system Hamiltonian. For a free trivalent RE ion, the Hamiltonian is as follows

$$H_{Isolation\ center} = H_E + H_{S.O.} = H_O + V + H_{S.O.} \tag{1.3}$$

H_E is the Hamiltonian of the electron system, and $H_{S.O.}$ refers to the most important magnetic interaction (spin-orbit interaction) of f electrons. H_O, as a main part of H_E, describes the interaction between each valence electron with the atomic kernel, which provides the electronic orbits of the 4f valence electrons forming the $(4f)^N$ electron configuration. V is the perturbation term, describing the Coulomb interactions among the 4f electrons. V makes the $(4f)^N$ electron configuration splitting to a series of states with different energy, each of which can be marked with L and S (the LS terms).

Generally, the mixed configuration induced by V is not obvious and can be ignored. For a large number of 4f electrons, L and S may represent more than one state, and an additional quantum number α is introduced.

The Hamiltonian $H_{S.O}$ describing the spin-orbit interaction, which can be understood as magnetic dipole–dipole interactions between the spin and angular moment of the electrons, can be expressed as follows,

$$H_{S.O.} = \sum_{i} \xi(r_i) l_i \cdot S_i \tag{1.4}$$

The spin-orbit coupling constant $\xi(r_i)$ is defined as a sole function of r_i, while r_i is the distance of i electron.

To construct wave functions for a multielectron atom on the basis of the central field approximation, Russell-Saunders or LS coupling scheme was chosen. It assumed that the matrix elements of $H_{S.O.}$ is smaller than the distances between different L (orbital moment) and S (spin moment), so the mixing of different L and S can be ignored. $H_{S.O.}$ can be written as $\zeta \boldsymbol{L} \cdot \boldsymbol{S}$, which separates L and S as states of series quantum numbers L, S, and J. J is the total angular momentum operator which has 2J + 1 eigenstates represented by the magnetic quantum number $M_J = -J, -J + 1, \ldots\ldots J$.

Actually, when different LS close each other, Russell-Saunders is not right. $H_{S.O.}$ will mix the states with the same quantum number J, M_J but differ LS, which results in the intermediate coupling eigenstates $f^N\{\alpha SL\}J>$ with the Russell-Saunders basis of ($|f^N \alpha SLJ>$).

$$\left|f^N\{\alpha SL\}J\right\rangle = \sum_{\alpha SL} c(\alpha SL)\left|f^N \alpha SLJ\right\rangle \tag{1.5}$$

$\{\alpha SL\}$ usually is the maximum contributor to the Russell-Saunders eigenstates.

For a RE ion in the crystal field, its Hamiltonian can be expressed as

$$H = H_{Isolation\,center} + H_{electron-static\,field} + H_{electron-lattice} + H_{lattice} \tag{1.6}$$

$H_{\text{electron-static field}}$ describes the interaction between electrons and static field. Ionic energy levels split under the influence of the static field produced by the crystalline environment, as shown in Fig. 1.4 which is the $4f$ energy levels of trivalent RE ions. Meanwhile, the interaction between ions with the static field will influence the free ion Hamiltonian, which results in the change of the whole energy levels.

$H_{\text{electron-lattice}}$ describes the influence of lattice waves on electrons. It will influence the light transition form and intensity and relates with the non-radiative relaxation.

H_{lattice} is the lattice vibration energy expressed by phonon mode.

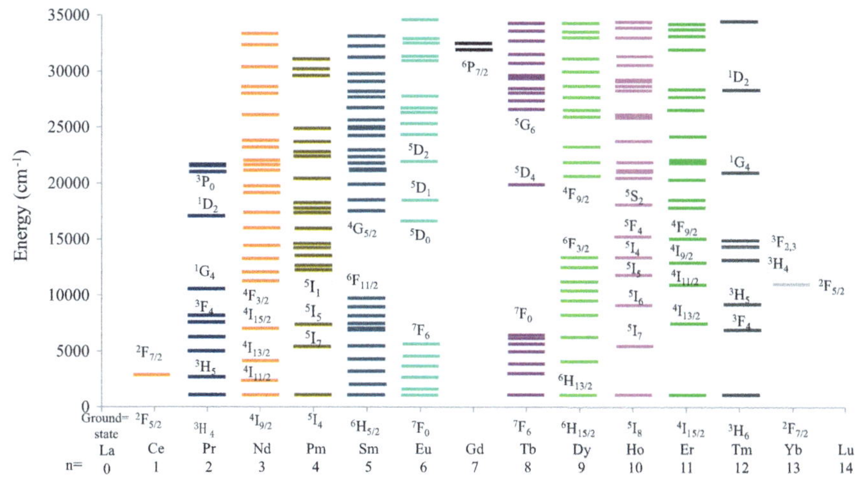

Fig. 1.4 Energy level of trivalent RE ions

1.2.2 The Radiative Transition of RE Ions

(1) **The basic principles on the transition** (Huang 2002; Yen and Selzer 1981)

The interaction operator of electrons with electric field or magnetic field is $-\mu\cdot\boldsymbol{E}$ or $-\boldsymbol{M}\cdot\boldsymbol{B}$, respectively, where $\mu = \sum_i e\boldsymbol{r}_i$, is the electric dipole moment. $M = \sum_i (e/2m)(\mathbf{l}_i + 2\mathbf{s}_i)$, is the magnetic dipole moment. The electron transition probability from the initial state $|i\rangle$ to the final state $|f\rangle$ is proportional to the matrix element of the operator between the state $|i\rangle$ and $|f\rangle$, which calls as the line strength S:

$$S(if) = \sum_{a,b} |\langle f|\mathbf{D}|i\rangle|^2 \tag{1.7}$$

D represents the electric dipole moment operator μ *or* the magnetic dipole moment operator M, the summation extends over all states of $|i\rangle$ and $|f\rangle$.

When the total spin of $|i\rangle$ state and $|f\rangle$ state is different, the matrix element is equal to zero. The spin selection rule for the electron transition is $\Delta S = 0$. However, the spin is not an effective quantum number due to the spin-orbit coupling, so the spin selection rule is not strictly followed. Generally, the spin allowed transition ($\Delta S = 0$) is stronger than the spin forbidden transition ($\Delta S \neq 0$).

In spectroscopy, another dimensionless parameter called the oscillator strength is defined as

$$f = \frac{1}{g_i} \sum \frac{8\pi^2}{3he^2} |\langle f|D|i\rangle|^2 \tag{1.8}$$

In this equation, m is the mass of an electron, v is the frequency of the transition, $hv = \left|E_f - E_i\right|$, g_i is the degeneracy of the state $|i\rangle$, for RE ions, it usually equals to $2J + 1$.

The light absorption intensity $I_\upsilon(l)$ is determined by measure the light absorption coefficient $\alpha(\upsilon)$:

$$I_\upsilon(l) = I_\upsilon(0)\exp[-\alpha(\upsilon)l] \tag{1.9}$$

$I_\upsilon(l)$ is the transmission intensity of light υ through the thickness l. The absorption cross section is defined as $\sigma(\upsilon) = \alpha(\upsilon)/N$, where N is the number of the luminescence center per volume. If the absorption is due to the electric dipole transition, there is the relationship between $\sigma(\nu)$, f_{ED} and S_{ED}:

$$\begin{aligned}\int \sigma(\upsilon)d\upsilon = N^{-1}\int \sigma(\upsilon)d\upsilon &= \frac{1}{4\pi\varepsilon_0}\frac{\pi e^2}{mc}\left[\left(\frac{n^2+2}{3}\right)^2 \cdot \frac{1}{n}\right] f_{ED} \\ &= \frac{1}{4\pi\varepsilon_0}\frac{8\pi^3\upsilon}{3hc}\left[\left(\frac{n^2+2}{3}\right)^2 \cdot \frac{1}{n}\right]\frac{1}{g_a} S_{ED}\end{aligned} \tag{1.10}$$

If the radiation transition is the electric dipole transition, the Einstein spontaneous emission rate A between the state $|i\rangle$ and $|f\rangle$ is:

$$\begin{aligned}A_{ED}(\upsilon) &= \frac{1}{4\pi\varepsilon_0}\frac{8\pi^2\upsilon^2e^2}{mc^3}\left[\left(\frac{n^2+2}{3}\right)^2 \cdot n\right] f_{ED} \\ &= \frac{1}{4\pi\varepsilon_0}\frac{64\pi^4\upsilon^2}{3hc^3}\left[\left(\frac{n^2+2}{3}\right)^2 \cdot n\right]\frac{1}{g_a} S_{ED}\end{aligned} \tag{1.11}$$

The above equation also fits to the magnetic dipole transition when substitute f_{ED} and S_{ED} by f_{MD} and S_{MD},$((n^2 + 2)/3)^2$ by n^2, respectively.

The relationship between the Einstein stimulated radiation rate B with the Einstein spontaneous emission rate A is:

$$A/B = (\hbar\omega^3/\pi^2c^3)n^3 = (4h\upsilon^3/c^3)n^3 \tag{1.12}$$

(2) **J–O theory**

J–O theory, established by B. R. Judd and G. S. Ofelt, is the only theoretical model to analyze the radiation transition of RE ions in solids (Song et al. 2016). It uses the integrated absorption intensity from the ground state to excited states in the measured absorption spectra to calculate the oscillator strengths of the electric dipole multipole transition, then to calculate the transition probabilities between different states.

Due to the screen from the $5s^2$, $5p^6$ orbits, 4f electrons of RE ions in solids are perturbed by the crystal field. Therefore, their spectra like that of free ions. The crystal field only effect the location of the energy level around several hundred cm^{-1}.

The energy level of all free RE ions is constituted by the same $(4f)^N$ configurations and has the same parity. Electric dipole transitions within $(4f)^N$ configurations are forbidden and magnetic dipole transition is allowed. When RE ions were doped in crystals, their luminescence were detected thanks to the contributions from configurations of opposite parity of $(4f)^{N-1}5d$ to the wave functions of the $(4f)^N$ configurations by the odd parity terms of the crystalline potential expansion. Therefore, the matrix elements of electric dipole operator between the mixed-parity eigenfunctions are nonzero. Then the equation of transition rates within $(4f)^N$ configurations was deduced by B. R. Judd and G. S. Ofelt starting from the point of mixed-parity configurations due to the static crystal field effect, which is called as J–O theory and is the only one theory to calculate the luminescence intensity of RE ions. Also it can be used in amorphous materials.

Adopting the static model, the system Hamiltonian should be $H = H_f + V$, H_f is the Hamiltonian of free ions, V is the crystalline potential and comes from the interaction between electrons and crystal field. Judd and Ofelt thought that the expression of the crystalline potential can be separated into even parity and odd parity parts. The odd parity parts caused the mixing of the opposite parity of $(4f)^{N-1}(n'l')$ configurations into $(4f)^N$ configuration, then electric-dipole transitions happened. Therefore, it required that the coefficient A_{tp} (t is odd values) is not equal completely to zero (Non central coordination field):

$$V = V_{even} + V_{odd} = \sum_{t,p} A_{tp} D_p^{(t)}$$

$$D_p^{(t)} = \sum_j^N r_j^t \left[\frac{4\pi}{(2t+1)}\right]^{1/2} \cdot Y_{t,p}(\theta_j, \phi_j) \tag{1.13}$$

where j represents j_{th} electron.

The wavefunction got by the diagonalization of the Hamiltonian of free ions and V_{even} was taken to be the zero-order approximate wave function, and the first-order wavefunction got by the consideration of the perturbation of V_{odd} as the crystal field eigenfunction, then the matrix element of the electric dipole operator $D_q^{(1)}$ between the crystal field eigenfunction should have nonzero values. The zero-order wavefunction was known if J mixing was not considered, as follows:

$$\langle A| = \sum \left\langle f^N \varphi J M \right| \cdot a_M \tag{1.14}$$

$\left\langle f^N \varphi J M \right|$ is the intermediate coupling state, a_M is the component coefficient, φ represents other quantum number. Furthermore, it can be written as the (S, L) coupled Russell-Saunders eigenstate $\left\langle f^N \varphi J M \right|$ according to the linear combination of *LS basis sets*.

According to the perturbation theory, the first-order wavefunction is

$$\langle B| = \langle A| - \sum_{\beta} \frac{\langle A|V_{odd}||\beta\rangle\langle\beta|}{E(A) - E(\beta)} \tag{1.15}$$

$\langle\beta|$ is the wavefunction of different parity excited configurations, $E(A)$ and $E(\beta)$ is the eigenvalues of $\langle A|$ and $\langle\beta|$, respectively.

The electric dipole moment operator P can be decomposed as $(P_q^{(1)})$ in the direction of x, y, and z, so there are transitions corresponding to various polarization. When $q = 0$, it is π polarization, $q = \pm 1$ corresponds to the σ polarization. The matrix elements of $P_q^{(1)}$ is not equal to zero due to the mixing of odd parity.

$$\langle B|P_q^{(1)}|B'\rangle = -\sum_{\beta} \frac{\langle A|V_{odd}|\beta\rangle\langle\beta|P_q^{(1)}|A'\rangle}{E(A) - E(\beta)} - \sum_{\beta} \frac{\langle A|P_q^{(1)}|\beta\rangle\langle\beta|V_{odd}|A'\rangle}{E(A') - E(\beta)} \tag{1.16}$$

Here the summation is for all $|\beta\rangle$ configurations with opposite parity. It is difficult to calculate above equation because that all $E(\beta)$, $|\beta\rangle$ and the odd components of crystal field contribute to the mixing of opposite parity configurations. At the same time in 1962, Judd and Ofelt independently made an important hypothesis to simply this calculation. They used a constant ΔE irrelevant with φ, J, M and β to substitute $E(A)$–$E(\beta)$ and $E(A')$–$E(\beta)$. It means that the energy splitting in the same configuration is far smaller than the energy difference of different configurations.

Therefore, to calculate above equation 1.16 we can get all J–O parameters. However, it is difficult to distinguish exactly the absorption lines corresponding to the transition between the components of two levels of the RE ions doped in crystal or amorphous materials. In experiment, it needs to sum the transition probabilities of all components of two levels, which is the line strength:

$$S_{jj'}^{ed} = \sum_{a,b} |\langle b|\mathbf{D}|a\rangle|^2 = e^2 \sum_{\lambda=2,4.6} \Omega_\lambda |\langle f^N\varphi J\,||U^{(\lambda)}||\,f^N\varphi' J'\rangle|^2 \tag{1.17}$$

$\langle\|U^{(\lambda)}\|\rangle$ is the reduced matrix element, which can be referred from the reference (Zhao et al. 2013) or other references. Ω_λ is not related to J, but only related to the crystal field parameters, so it can be used as the adjusted parameter.

1.2.3 The Upconversion Luminescence Based on Excited State Absorption Process

(1) **The excited state absorption (ESA) process**

ESA governed UCL was first proposed by Bloembergen et al. In a typical ESA process (Hehlen et al. 1996), the electrons are pumped by the incident light to the excited state, then to a higher energy state by further absorbing incident photons, which results photons with high energy emission. As shown in Fig. 1.5 of the ESA

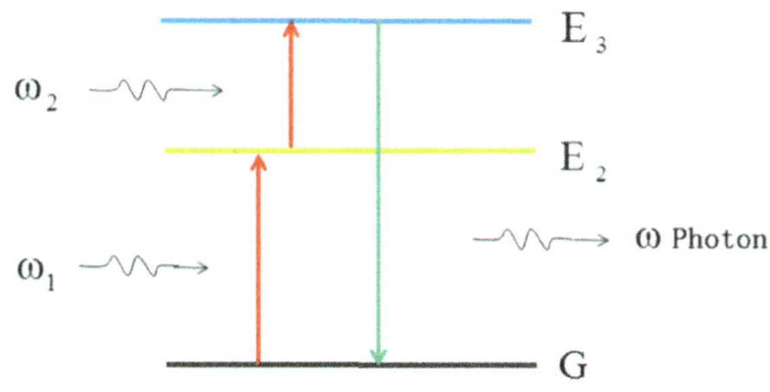

Fig. 1.5 Excited state absorption process of upconversion

process, the electron at the ground state (G) of the luminescence center absorbs a photon with the frequency of ω_1 and is excited to the metastable excited state E_2. Subsequently, the electron at E_2 absorbs another photon with the frequency of ω_2 and is excited to the higher excited state E_3. The electron transition from E_3 to G gives a photon with the frequency of ω. It is clear that $\omega > \omega_1, \omega_2$. Hence, the UCL is achieved.

The electron transition in the ESA process can be either direct transition or phonon-assisted transition. Generally, the ESA process requires two pump lasers, one of which resonates with the ground state absorption (GSA) and the other one resonates with the excited state absorption. In some conditions, one laser pump may also be possible, including (i) the absorption energy of the ground state and the excited state are uniform or the absorption energy difference can be compensated by the phonons, and (ii) the transitions are in homogeneously broadened in some irregular materials. The ESA process occurs inner one ion, and hence, it is not affected by the concentration of the RE ions which avoids the losses induced by the energy transfer between RE ions.

(2) **Examples of the UCL based on ESA process** (Xu and Su 2004)

With the pump by Kr^+ ion laser at 647.1 nm, the luminescence from the excited states $^3H_4$, $^1G_4$, $^1D_2$, and $^1I_6$ of $LaF_3{:}Tm^{3+}$ can be observed, as shown in Fig. 1.6. The UCL from the higher excited states ($^1G_4$, $^1D_2$ or $^1I_6$) is caused by the ESA process. The excitation steps are as follows. The electron absorbs the first photon and jumps from the ground state $^3H_6$ to the excited state $^3F_2$. As the states $^3F_2$, $^3F_3$, and $^3H_4$ are very close, the electron quickly relaxes to the state $^3H_4$. The electron may absorb another photon and jumps from the state $^3H_4$ to $^1D_2$, or backs to the ground state by giving an infrared photon. The electron at the state $^1D_2$ may back to the ground state or the state $^3F_4$ by giving a visible photon. The electron at the state $^3F_4$ may absorb other one or two photons and jumps to much higher states ($^1G_4$ or $^3P_1$). As the states $^3P_1$ and $^1I_6$ are very close, so the electron at $^3P_1$ will quickly relax to $^1I_6$. Therefore, the UCL can be observed from the excited states $^1G_4$, $^1D_2$, and $^1I_6$.

Fig. 1.6 Energy level of Tm^{3+} in LaF_3 and upconversion process under the excitation of 647.1 nm

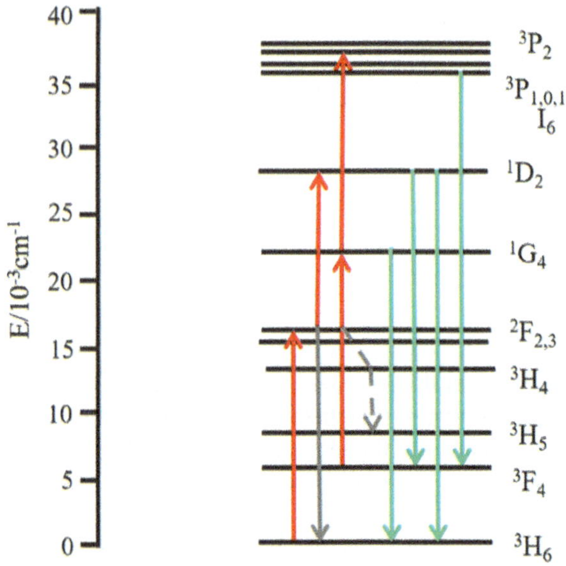

1.2.4 The Upconversion Luminescence Based on Photon Avalanche

(1) **The photon avalanche (PA) process**

The photon avalanche (PA, or absorption avalanche) process was firstly discovered in $LaCl_3$:Pr^{3+} material system (Chivian et al. 1979). In this system, the pump frequency is resonant to the excited state absorption from $^3H_5$ to $^3P_1$ state of Pr^{3+} ion. Beyond a critical pump intensity, the luminescence from $^3P_1$ or $^3P_0$ state of Pr^{3+} ion is enhanced sharply, and accordingly the absorption of the pump is also increased sharply. This is the PA phenomenon. The PA process was also revealed in other material systems like $LaBr_3$:Sm^{3+} (Krasutsky 1983) and $CeCl_3$:Nd^{3+} (Pelletier-Allard and Pelletier 1991). As PA produces population in an excited state with energy exceeding that of the pump photon, it is possible to realize UCL.

The photon avalanche process combines the excited state absorption and the energy transfer processes. As shown in Fig. 1.7, it shows the condition in a four-level system, where G, m_1, and m_2 are the ground state and two metastable states, respectively, E is the excited state emitting photons. The pump laser with the frequency of ω' enables the resonant absorption from m_1 to E, which is a characteristic of PA process. Though the ground state absorption may be not resonant to the pump, a small amount of electrons may absorb the pump photons and jumps to the excited states between E and m_2, then relax to m_2. The electron at m_2 jumps to m_1 and simultaneously transfers the energy to an electron at the ground state G to m_1. Meanwhile, the electron transition from E to m_1 may excite two electrons from G to m_1. Hence, the resonant pumping of electrons from m_1 to E makes the electron population at E

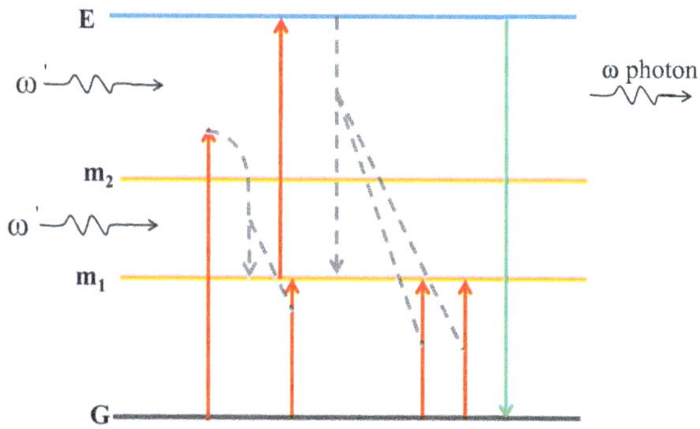

Fig. 1.7 Photon avalanche process

increasing like avalanche. The electron transition from E to G gives a photon with a higher frequency of ω (ω > ω′), leading to the UCL.

Another characteristic of PA process is an excitation power threshold (Kueny et al. 1989; Guy et al. 1994), which divides the excitation into two different stages. In the condition of a weaker excitation than the threshold, the UCL is quite weak and the crystal is transparent for the pump. In the condition of a stronger excitation than the threshold, the intensity of the UCL is enhanced by several orders and the absorption of the pump light is strong. Moreover, the concentration of the RE ions shall be high enough to enable the population of the metastable states by energy transfer.

The UC laser based on PA process avoids the photo-induced degradation of host materials observed in the laser based on the down-conversion emission by violet light excitation. Compared with the laser based on ESA process, the laser based on PA process only requires one resonant transition, it means only pump light needed. Hence, PA process is also an ideal pump scheme for the UC laser with high dopant concentration. More information on the mechanisms and the application in lasers of the PA process can be found in the review of 《*Photon avalanche upconversion in rare earth laser materials*》 (Joubert 1999).

(2) **Example of the UCL based on PA process**

In $CdF_2:Tm^{3+}$ system, the PA-induced UC emission, which converts the red excitation light to the blue light, can be observed at 77K. The pump light with a wavelength of 647.7 nm is resonant to the $^3H_4$–$^1D_2$ transition from Tm^{3+} ions, which produces the blue emission peaking around 450 nm (Ofelt 1962). In $LiYF_4:Tm^{3+}$ system, the PA-induced UC emission is also proved, which converts the NIR excitation light to the UV light (Kueny et al. 1993). In this system, the pump photons with a wavelength of 1.04 μm enables the avalanche excitation of $^3F_4$–$^3F_2$ transition, which populates the energy level with a wavenumber of 15,000 cm^{-1}; subsequently, the population of the high energy state with a wavenumber of 35,000 cm^{-1} is realized by multiple energy transfer processes, leading to the UV emission.

1.2.5 The Upconversion Luminescence Based on Energy Transfer

(1) **The mechanism of the energy transfer upconversion process**

The energy transfer can occur between either same kind or different kinds of RE ions. Thus, the UCL based on the energy transfer includes two different types (Xu et al. 2004), (i) the energy transfer between the same kind of ions at the excited states, and (ii) the sequential energy transfer between different kinds of ions (the generally defined energy transfer upconversion (ETU) process). Similar to the definitions in the phosphor field, the ions which absorb light and give the energy are called as the sensitizer (or donor), while the ions accepting the energy and giving the photons are called as the activator (or acceptor).

A. **Basic principles involved in the energy transfer processes**

There are four types of energy transfer process that include the resonant radiative energy transfer, the resonant non-radiative energy transfer, phonon-assisted non-radiative energy transfer, and the cross-relaxation between the same kind of ions. In the resonant radiative energy transfer, the sensitizer emits a photon, followed by the photon absorption by the activator. This type of energy transfer depends on the shape of the sample, and also the overlap between the emission spectrum of the sensitizer and the absorption spectrum of the activator. The resonant radiative energy transfer is rarely employed in the UCL.

The resonant non-radiative energy transfer, i.e., the Föster energy transfer, is the most important type in UCL. In this scheme, the sensitizer ion at its excited state transfers its energy to the acceptor ion through the Coulomb interactions before it is able to emit a photon, and the overlap of the wave functions between the two ions are not essential. Föster firstly considered this type of energy transfer as dipole–dipole interactions (Föster 1948). He assumed that the interaction is strongest in the condition that the electric dipole transitions are all permitted in these two ions (Auzel 1973). The transfer probability is given by:

$$P_{SA} = \frac{2\pi}{\hbar} \left|\langle S^e A^0 | H_{SA} | S^0 A^e \rangle\right|^2 \rho_E \tag{1.18}$$

S, A the sensitizer center and the activator center, respectively
H_{AS} interaction Hamiltonian
e, o the excited state and the ground state, respectively
ρ_E density of states provided by the vibrational motion contributing to the line broadening of the transition.

For dipole–dipole interactions, the transfer probability can be written as

$$P_{SA} = (R_0/R)^i / \tau_s \tag{1.19}$$

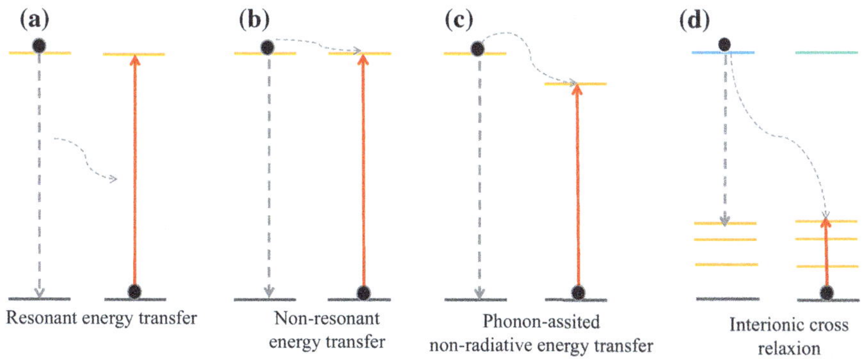

Fig. 1.8 Different energy transfer processes between two ions

τ_s actual lifetime of the sensitizer excited state
R_0 critical transfer distance for which excitation transfer and spontaneous deactivation of the sensitizer have equal probability.
R the actual distance between the sensitizer and the activator, the transfer probability decreases sharply for $R > R_0$.

The resonant energy transfer shown in Fig. 1.8a can occur if the sensitizer center S and the activator center A satisfy two conditions: one is that the energy difference between the excited state and the ground state are the same for the two ions, and the other one is that the distance between the ions is close enough. If we consider two ions with different excited state energy, i.e., the energy mismatch exists between the sensitizer center S and the activator center A, the resonant energy transfer can not occur. However, it is experimentally found that the energy transfer between S and A can take place assisted by the phonons, i.e., the phonon-assisted non-radiative energy transfer, as shown in Fig. 1.8c. In this type of energy transfer, the transfer probability is given by:

$$W_{nr} = W_0 \exp(-\Delta E/\hbar\omega) \tag{1.20}$$

W_0 the transfer probability in the condition of zero energy mismatch
ΔE *the mismatch energy between the sensitizer center S and the activator center A*
$\hbar\omega$ the phonon energy of the host material.

In energy transfer between RE ions, energy mismatch as high as several thousand reciprocal centimeters are encountered, multiphoton assisted energy transfer have to be considered.

B. **Energy transfer upconversion (ETU) process**

Before 1966, all energy transfer considered was that the sensitizer ion is in one of its excited state while the activator is in its ground state, as depicted in Fig. 1.9a. Auzel proposed that the activator ions at their excited states were also able to accept the

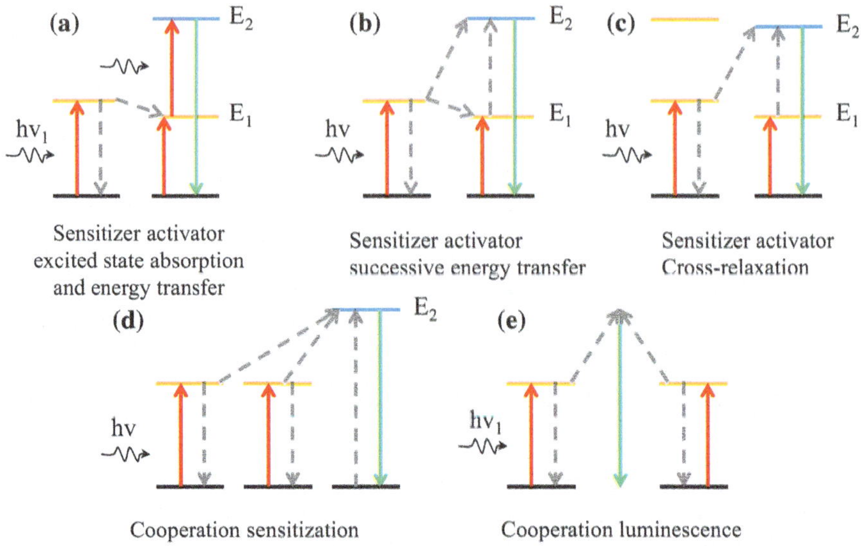

Fig. 1.9 The mechanisms of energy transfer (Malinowski et al. 2000)

energy from sensitizer ions (Auzel 1966), as shown in Fig. 1.9b. This kind of energy transfer is also defined as the sequential energy transfer. The transfer probability can then be rewritten as:

$$P_{SA} = \frac{2\pi}{\hbar} \left| \langle S^e A^0 | H_{SA} | \, S^0 A^{ee} \rangle \right|^2 \rho_E \tag{1.21}$$

where $|S^0 A^{ee}\rangle$ the wave function for the system where the sensitizer is in its ground state and the activator in a doubly excited state.

The energy transfer from the sensitizer to the activator decreases the electrons and lifetime of the excited state of the sensitizer, resulting in the weak emission or the absence of the luminesce from the sensitizer ions. As the excited state of the sensitizer is excited by a photon with low energy, the energy transfer enables the activator ion to jump to its high excited states of which the energy may be higher double, triple or even quadruple than that of the excitation photons. When the activator ion jumps from the high excited state to its lower state, it will emit a photon with high energy. As the activator contains several excited states and only one ground state, the electron population of higher excited states will be achieved by the energy transfer through the energy superposition of several pump photons. This implies the probability of the multiphoton upconversion. Thereby, people understood that the energy difference between different levels, instead of the absolute energy of the energy level, is more important to the energy transfer between ions. The energy transfer between the same kind of ions is also called the cross-relaxation upconversion, as shown in Fig. 1.9c

Besides the successive energy transfer described above, the cooperative effect of ions by their mutual interaction also be used to explain anti-Stokes luminescence, such as the cooperative sensitization (Fig. 1.9d) and cooperative luminescence (Fig. 1.9e). In a cooperative sensitization process as can be seen from Fig. 1.9e, the united two excited sensitizer ions transfer their energy to an activator ion meanwhile, which enable the acceptor ion to jumps to its high excited energy level. The cooperative luminescence is the emission in a single process of one photon from two excited interacting ions.

All of the above energy transfer processes are due to the interactions between the RE ions, so they depend on the concentrations of the RE ions. If the energy between the sensitizer and the activator is mismatched in the energy transfer process, the energy difference can be compensated by the phonon. In practice, the energy difference of the different energy transfer process can be used to estimate which energy transfer process is the most probable. The outstanding advantage of utilizing energy transfer process to realize the upconversion laser is that it only requires single pump with a certain frequency.

(2) **The examples based on ETU process** (Xu and Su 2004)

The UCL of Pr^{3+} ion based on ETU process. Upon the pump of two lasers with the wavelength of 1017 nm and 835 nm, respectively, Pr^{3+} ions doped in fluoride host can emit visible lights at different wavelengths. As shown in Fig. 1.10 of the upconversion processes of Pr^{3+} in fluoride glass LaF_3, the excitation processes are as follows. The electron at ground state $^3H_4$ absorbs a photon with the wavelength of 1017 nm and jumps to the excited state $^1G_4$, followed by jumping to upper excited states $^3P_0$, $^3P_1$, and $^1I_6$ upon excitation at 835 nm. The transitions from $^3P_0$, $^3P_1$, and

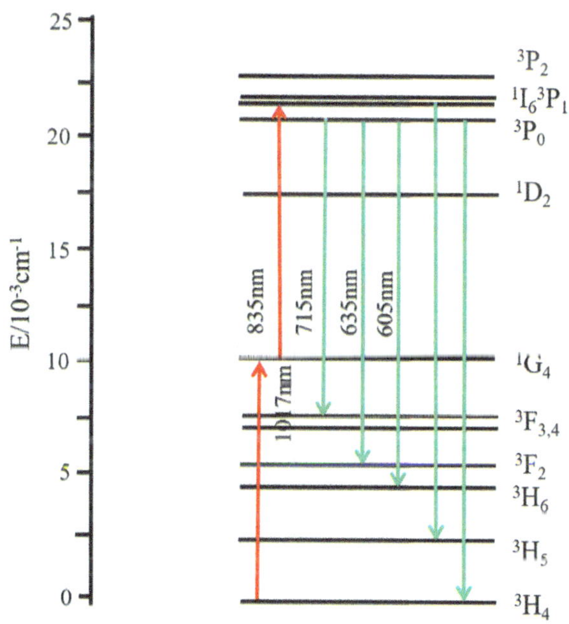

Fig. 1.10 Upconversion processes of Pr^{3+} in fluoride glass LaF_3

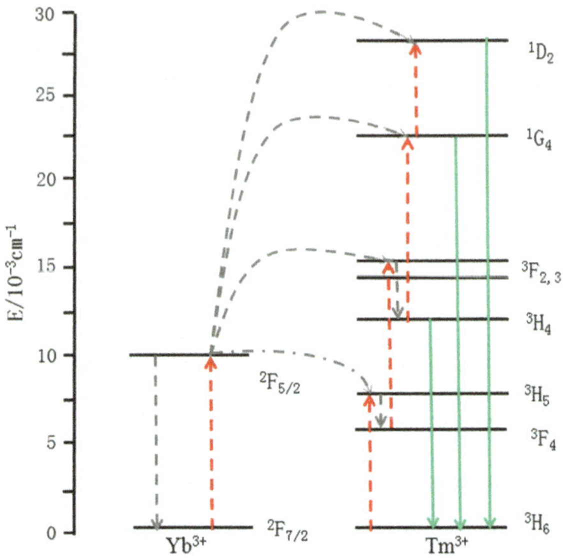

Fig. 1.11 Upconversion processes of Yb^{3+}–Tm^{3+} through successive energy transfer

$^1I_6$ states to different lower excited states exhibit the UCL at different wavelengths, as can be seen in Fig. 1.10.

The UCL from Yb^{3+}–Tm^{3+} system based on ETU process. Figure 1.11 schematically illustrates the UC process based on the energy transfer from Yb^{3+} ions to Tm^{3+} ions. Yb^{3+} ions are excited by an infrared light with a wavelength of 980 nm, accompanied by the electron transition from the ground state $^2F_{7/2}$ to the excited state $^2F_{5/2}$, which transfer the energy to the neighboring Tm^{3+} ions and back to the ground states. The Tm^{3+} ions are excited from the ground state $^3H_6$ to the excited state $^3H_5$. Afterward, the Tm^{3+} ions at the excited state $^3H_5$ are further excited to the upper excited states by absorbing the energy transferred from neighboring Yb^{3+} ions. The Tm^{3+} ions at the upper excited states back to the ground state and emit the UCL.

1.2.6 The Differences Among the Upconversion Luminescence Processes

In the UCL process, one or more UC process may be involved. For example, in Yb–Er co-doped system, ETU and ESA probably simultaneously exist in the UCL process. The sequential absorption in the ESA process is due to the combination absorption by many excited states of the activator ions, whereas in the ETU process is due to the absorption by the ground state of the sensitizer ions (the excited state absorption relies on the energy transfer). Hence, in ETU process, the excitation spectra derive from the sensitizer ions, and this feature can be utilized to distinguish whether ETU or ESA is the main processes creating UCL (Nadort et al. 2016). For instance, in $NaYF_4$:2%Er, 18%Yb system (Suyver et al. 2005), only the transition $^2F_{7/2} \rightarrow {}^2F_{5/2}$ of Yb^{3+} ions is observed in the excitation spectrum, which means that the UCL of this system is mainly due to ETU from the sensitizer Yb^{3+} ions to the activator $Er3^+$ ions. In $NaYF_4$:2%Er system (Suyver et al. 2005), the excitation spectrum contains many transitions such as $^4I_{15/2} \rightarrow {}^4I_{1/2}$ and $^4I_{11/2} \rightarrow {}^4F_{7/2}$, which implies that ESA is the main process for the sequential absorption of photons by Er^{3+} ions.

The light absorption rate in different upconversion processes is also different. In ESA UC process, the absorption by the ground state and metastable excited states occurs in nanosecond time scale (Zhao et al. 2016), and thus, the time to UCL is also fast. The corresponding transient photoluminescence (PL) decays mono-exponentially with the elapsing time, as shown in Fig. 1.12a. In ETU process, it requires a long time as the luminescence of the activator can be produced under at least twice energy transfer procedure. In the related transient PL, the PL intensity firstly increases with the elapsing time (generally in the time scale of microseconds) and then decreases. Hence, in the decay profile of the UCL governed by ETU process, the decay rate of the sensitizer, the rate of the energy transfer and the PL decay rate of the activator are all included. In the material system in which simultaneously exist two kinds of UC processes, the PL intensity of the activator decays bi-exponentially (Fig. 1.12b).

The PA process features an excitation threshold, so its existence can be evaluated by the relation between the luminescence intensity and the excitation power. The UCL intensity I generally changes exponentially with the excitation power P, as shown in Fig. 1.13a (Sangeetha and van Veggel 2009), and no threshold power can be observed. In contrast, in PA process, the intense UCL can only be observed upon a certain excitation power. As can be seen from Fig. 1.13b (Singh et al. 2011), which shows the UCL based on the PA process from $LaF_3:Er^{3+}/Yb^{3+}$ system. It is clear that in condition of a lower excitation power than the threshold, few UCL can be observed, while the non-radiative and the downconversion processes govern the transitions. Upon the threshold, the UCL intensity is increased sharply, indicating

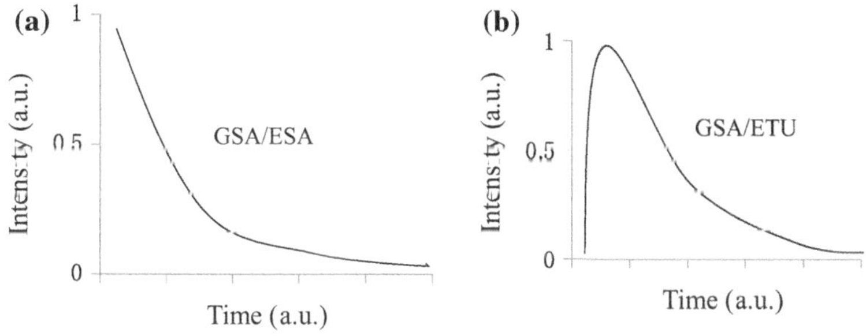

Fig. 1.12 The evolution of upconversion luminescence over time after the excitation from **a** GSA/ESA and **b** GSA/ETU processes

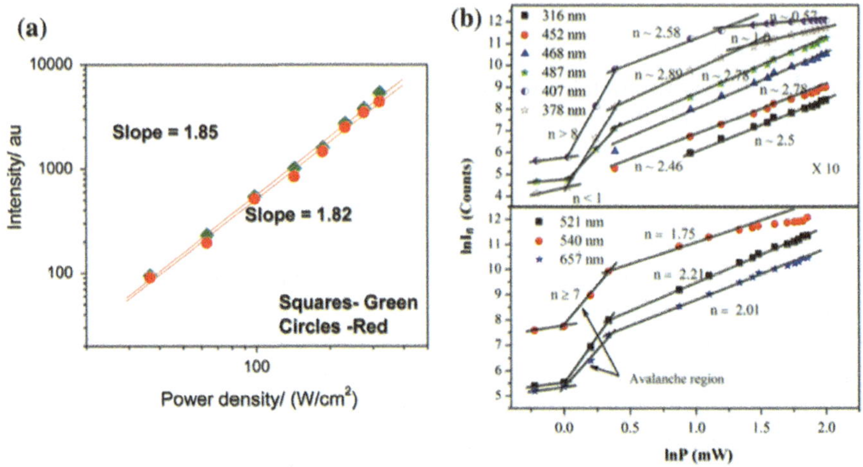

Fig. 1.13 **a** Variation of upconversion emission intensity with excitation power density for lanthanum zirconate doped with a Ho^{3+}, Yb^{3+} ratio of 1:3 and 1:2, respectively (Sangeetha and van Veggel 2009). **b** Variation of upconversion emission intensity with excitation power density for various bands from LaF_3:Er^{3+}/Yb^{3+} (Singh et al. 2011)

that the PA process can only occur in the condition of the excitation pump power beyond the threshold.

The ETU process is likely to be confused with the cooperative sensitization energy transfer UC process. For two-photon process, the luminescence intensity from both processes changes proportionally with the square intensity of the excitation power. If the lifetime of the emissive energy state of activator ions is much shorter than that of the excited energy state of sensitizer ions, then the UCL lifetime of these two processes equals to the half-lifetime of the excited energy state of sensitizer ions. Generally, these two processes coexist while the probability of ETU process is much larger (Auzel 1990). Only in some special cases like the absence of real energy states for energy transfer, such as in Yb^{3+}–Tb^{3+} system (Kaplyanskii 1987), the cooperative sensitization process plays the main role in the UCL. In addition, the probability of the cooperative sensitization process will larger in ion clusters.

Another kind of the UC process is the cooperative luminescence, which mainly be observed in some crystals with special crystal structure, such as Yb^{3+} ions doped $CsCdBr_3$ (Hehlen et al. 1999; Goldner et al. 1997), $Cs_3Y_2Br_9$ (Lüthi et al. 1998) and $Cs_3Lu_2Br_9$ (Hehlen et al. 1996). In these crystals, a large proportion of Yb^{3+} ions are paired, and the cooperative luminescence is produced when two excited Yb^{3+} ions give the energy out simultaneously.

1.3 Managing the Upconversion Luminescence Color and Efficiency of Rare Earth Ions

1.3.1 Managing the Upconversion Luminescence Color of Rare Earth Ions

The management of the UCL color of RE ions is critical and essential for their applications in many fields. This is especially important in the complex conditions in the biological diagnosis field, which requires the UCL to be highly recognized. The UCL color of RE ions is effected by many factors, such as the type and the concentration of the dopant ions, the type of the host, and the fabrication methods (Hutchinson and Allik 1992). The color of the RE ions is commonly adjusted through the dopant ions, the host and the size of the UC nanoparticles.

(1) **Managing the emission color through the dopant ions**

The most efficient method to change the luminescence color is choosing different RE ions, as each type of RE ions has its unique emission peak. Table 1.2 shows the luminescence properties of the typical doping systems of the UC nanoparticles. In these systems, Yb^{3+} is contained as the sensitizer to absorb the excitation light. From Table 1.2, it is clear that different activators yield different light color, enabling the

Table 1.2 Dopants and hosts for multicolor upconversion crystal (Heer et al. 2003, 2004; Yi and Chow 2006; Liu and Chen 2007; Qin et al. 2007)

		Luminescence			
Doped ions	Host	Blue	Green	Red	References
Tm^{3+}	α–$NaYF_4$	450, 475 (S)		647 (W)	Yi and Chow (2006)
	β–$NaYF_4$	450, 475 (S)			Liu and Chen (2007)
	LaF_3	475 (S)			Heer et al. (2003)
	$LuPO_4$	475 (S)		649 (S)	Qin et al. (2007)
Er^{3+}	α–$NaYF_4$	411 (W)	540 (M)	660 (S)	Yi and Chow (2006)
	β–$NaYF_4$		523, 542 (S)	656 (M)	Liu and Chen (2007)
	LaF_3		520, 545 (S)	659 (S)	Heer et al. (2003)
	$YbPO_4$		526, 550 (S)	657, 667 (S)	Qin et al. (2007)
	$Y2O_3$		524, 549 (W)	663, 673 (S)	Wang and Liu (2009)
Ho^{3+}	α–$NaYF_4$		540 (S)		Wang and Liu (2009)
	LaF_3		512 (S)	615, 658 (M)	Heer et al. (2003)
	$Y2O_3$		543 (S)	665 (M)	Wang and Liu (2009)

luminescence color changing from blue to red (Heer et al. 2003, 2004; Yi and Chow 2006; Liu and Chen 2007; Qin et al. 2007).

In the multiple doping systems, the luminescence color can be tuned either by varying the type or the concentration of the doping ions. The doping concentration governs the number and the distance of the doping ions in the host matrix, which is important for the UC luminescence properties. For example, Liu et al. (Wang and Liu 2009) utilities the multiple ions co-doping method, and through varying the doping concentration of different ions to realize the emission of many colors under a single wavelength excitation (980 nm). By changing the doping concentration of the two kinds of ions, $NaYF_4$ nanoparticles co-doped Yb/Er with a concentration ratio of 18/2 mol% present sharp emission peaks from $^2H_{9/2}$–$^4I_{15/2}$, $^2H_{11/2}$, $^4S_{3/2}$–$^4I_{15/2}$ and $^4F_{9/2}$–$^4I_{15/2}$ transitions of the Er^{3+} ions (Fig. 1.14a). These peaks correspond to

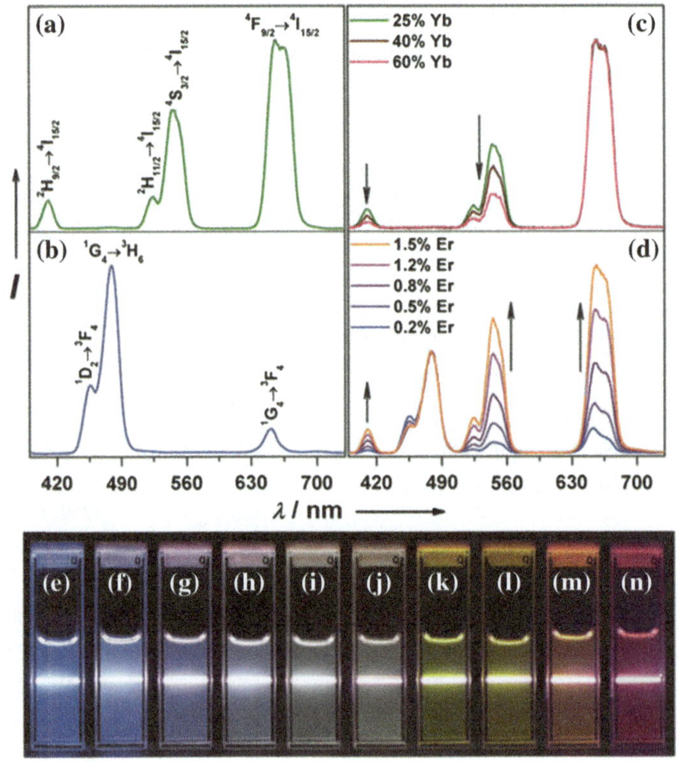

Fig. 1.14 Upconversion luminescence spectra of nanoparticles in ethanol solution (10 mM) at room temperature. **a** $NaYF_4:Yb^{3+}$, Er^{3+} (18/2 mol%), **b** $NaYF_4:Yb^{3+}$, Tm^{3+} (20/0.2 mol%), **c** $NaYF_4:Yb^{3+}$, Er^{3+} (25–60/2 mol%) and **d** $NaYF_4:Yb^{3+}$, Tm^{3+}, Er^{3+} (20/0.2/0.2–1.5 mol%). In **c** and **d**, the spectra are normalized by the emissions of Er^{3+} (660 nm) and Tm^{3+} (480 nm), respectively. **e**–**n** Photos of the colloidal solution under the excitation of 980 nm. **e** $NaYF_4:Yb^{3+}$, Tm^{3+} (20/0.2 mol%), **f**–**j** $NaYF_4:Yb^{3+}$, Tm^{3+}, Er^{3+} (20/0.2/0.2–1.5 mol%) and **k**–**n** $NaYF_4:Yb^{3+}$, Er^{3+} (18–60/2 mol%) (Wang and Liu 2009)

blue, green, and red emissions, respectively, making the nanoparticles displaying the orange color (Fig. 1.14k). Increasing the doping concentration of Yb^{3+} ions decreases the distance between Yb–Er ions, which promotes the back energy transfer from Er^{3+} to Yb^{3+} ions and suppress the excitations of $^2H_{9/2}$, $^2H_{11/2}$ and $^4S_{3/2}$ levels, leading to the reduced intensity of the blue ($^2H_{9/2}$–$^4I_{15/2}$) and green ($^2H_{11/2}$, $^4S_{3/2}$–$^4I_{15/2}$) emissions (Fig. 1.14c). It shows that the relative intensity of the three colors can be precisely adjusted (as shown in Fig. 1.14i–n) and the multiple emission colors can be obtained by increasing the doping concentration Yb^{3+} ions.

By utilizing dual emission process through doping three kinds of RE ions, the color of the UC emission in the visible region can also be managed. Through adding two kinds of emission ions (Tm^{3+} and Er^{3+}) and adjusting their concentrations, the relative emission intensity of these two kinds of ions can be precisely managed (Fig. 1.14d), and thus the emission color can be tuned from blue to white (Fig. 1.14 f–j).

(2) **Influence of the host materials**

As the dopant ions in different host lattice have different symmetry, the emission color will also be different. From Table 1.2, it shows that the RE ions exhibit different emission colors in different host materials.

In the system containing several kinds of dopant ions, the emission color can be different in different hosts despite of the same doping concentration. For example, in Ho^{3+} and Yb^{3+} co-doped system (Sangeetha and van Veggel 2009), as shown in Fig. 1.15, the green emission locating around 550 nm deriving from the transition from $^5S_2$ to $^5I_8$ of Ho^{3+} ions is more intense in zirconate host than that in silicate host. This may correlate with the low phonon energy of the zirconate host, which reduces the multiphonon relaxation process of $Ho^{3+}(^5I_6) \rightarrow Ho^{3+}(^5I_7)$ (corresponding to red emission), leading to the enhanced green emission and the reduced red emission.

(3) **Influence of the nanoparticle size**

The UC color of the UCPs can also be adjusted by their size. For instance, Capobianco et al. (Vetrone et al. 2004) found that the red emission from Y_2O_3:Yb/Er

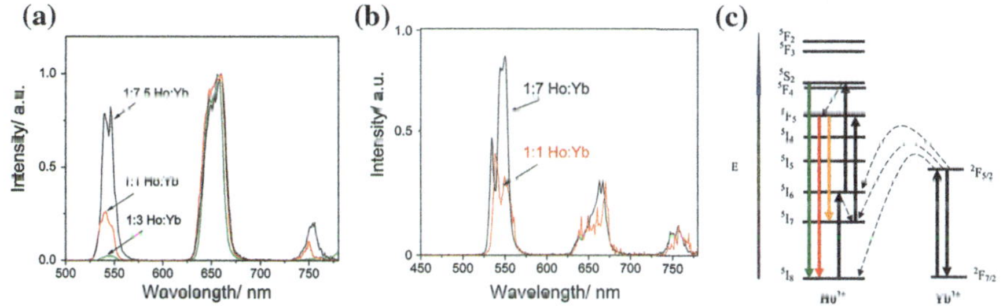

Fig. 1.15 Normalized upconversion emissions spectra of Yb^{3+}, Ho^{3+} co-doped silicate compound (**a**) and zirconate (**b**). **c** Schematic illustration for the mechanism of upconversion (Sangeetha and van Veggel 2009)

UCPs was enhanced with the UCP diameter of 20 nm. Song et al. (Wang et al. 2010) studied the luminescence properties of the Y_2O_3:Yb/Er UCPs with different diameters (13–55 nm), which showed that the relative intensity of blue and green emission could be adjusted by varying the UCPs size. In $NaGdF_4$:Yb/Tm UCPs, cutting the size of the UCPs from 25 to 15 nm reduces the relative intensity of blue emission ($^1G_4$–$^3H_6$, 480 nm) compared to that of near-infrared emission ($^3H_4$–$^3H_6$, 800 nm), as shown in Fig. 1.16, leading to the change in the emission color (Wang et al. 2010).

As the exciton Bohr radius of the Ln-doped UCPs is quite small, hence, the mechanism for the management of the emission color by the size of UCPs will be different. It can be attributed to the surface effect (Li et al. 2012), other than the quantum size effect like in the quantum dots. Minishing the size of UCPs, the relative concentration of the surface doping ions increases. As the emission color is the sum of the emission from both surface and inner ions, hence, varying the UCP size can adjust the emission color by changing the relative intensity of the emission from surface and inner ions.

(4) **Influence of the excitation conditions**

As different RE ions show different response to the excitation light, hence, the emission color of the UCPs with multiple ions can also be tuned by changing the excitation conditions. Deng et al. (2015) reported the emission property of the core/shell UCPs with $NaYF_4$ as the host. Under the laser excitation at 808 nm, Yb^{3+} ions are excited and transfer their energy to Tm^{3+}, leading to the blue emission from the first shell. Under the excitation at 980 nm, the Yb/Ho/Ce system is excited, leading to the red emission from Ho^{3+}. Therefore, combining with the change in the excitation conditions, multicolors can also be obtained from the UCPs.

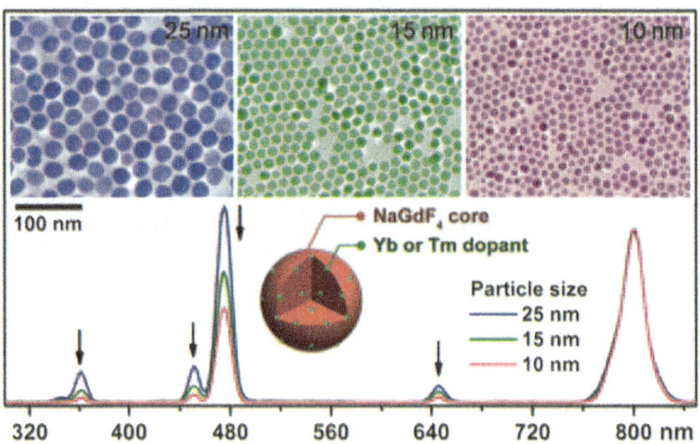

Fig. 1.16 Normalized upconversion luminescence spectra and TEM images of NaGdF4:Yb^{3+}, Tm^{3+} (25/0.3 mol%) nanoparticles with different sizes in cyclohexane solution, the wavelength of excitation is 980 nm and the power density is 10 Wcm^{-2} (Wang et al. 2010)

1.3.2 *Optimization in Luminescence Efficiency and Brightness of the UCPs*

(1) **Effect of the core/shell structure**

In most cases, the lower efficiency of UCNPs, in comparison with their bulk counterparts, is due to the increased non-radiative recombination induced by the surface defects. As smaller UCNP possesses larger specific surface area, increasing the size of UCNPs can enhance the luminescence efficiency. However, this enhancement is little effective, as larger UCNPs will limit their application in many fields. Hence, a practical method to enhance the luminescence efficiency is to utilize core/shell structure. Many core/shell structures, such as $NaYF_4$:Yb, Er/Tm@$NaYF_4$ (Liu et al. 2009), $NaYF_4$:Yb, Er@$NaGdF_4$ (Guo et al. 2010), $NaGdF_4$:Yb, Tm@$NaGdF_4$ (Park et al. 2009), KYF_4:Yb, Er@KYF_4 (Schäfer et al. 2008), and YOF:Yb, Er@YOF (Yi et al. 2011), are proved to be able to enhance the UCL efficiency. In the core/shell structures, the shell materials are typically similar to the host materials of the core. Compared to the uncoated UCNPs, the core/shell structured UCNPs possess higher UCL intensity. By doping Ln-ions into the shell, the UCL intensity can be further increased, such as in the β-$NaYF_4$:Yb, Tm@β-$NaYF_4$:Yb, Er core/shell structure (Qian and Zhang 2008).

(2) **Effect of the host material**

The host lattice manipulation may vary the crystal field asymmetries around the emitting Ln-ions, and higher asymmetry can increase the probability of ED transitions and radiative recombination. It is found that the enhanced asymmetry of the crystal field around Er^{3+} ions, by applying an external electric field on $BaTiO_3$: Yb/Er film, enables the enhanced green emission by a factor of 2.5 (Hao et al. 2011), which proves the function of the crystal field asymmetry in affecting the luminescence intensity. Doping small ions into the host lattice, such as Li^+ ions, can change the crystal lattice constant and the crystal field asymmetries, which increases the UCL intensity by 1–2 orders (Chen et al. 2008).

(3) **Other methods to improve the UCL efficiency**

Surface plasmon-coupled emission (SPCE). By anchoring Ag or Au nanostructures on the surface of the UCNPs, the UC emission intensity can be enhanced. This is due to the enhanced radiative recombination rate by the surface plasmon induced by the metal nanostructures. For example, Yan et al. (Feng et al. 2009) firstly reported the enhanced UC emission from $NaYF_4$:Yb, Er UCNPs by coupling Ag nanowires, and the enhancement of the red emission (650 nm) is much stronger than the green emission (550 nm). The UC emission of $NaYF_4$:Yb, Tm UCNPs is enhanced by a factor of 8 through coupling Au shell (Zhou et al. 2012).

Dye antenna enhanced UC emission. It is found that the UC emission can also be enhanced by attaching dye molecules on the surface of the UCNPs. The dye molecules act as antenna for light absorption, which can increase the absorption section and the absorption bandwidth, leading to the increased UC emission efficiency. For example, Schuck et al. (Garfield 2018) utilize dye IR806, which is adsorbed at the surface of $NaGdF_4$ matrix, to enhance the UC emission from Yb^{3+} based UCNPs with a 33,000-fold increase in brightness and a 100-fold increase in efficiency over bare UCNPs.

References

Auzel F. Compteur quantique par transfert d'energie entre deux ions de terres rares dans un tungstate mixte et dans un verre. CR Acad Sci Paris. 1966;262:1016–9.

Auzel FE. Materials and devices using double-pumped-phosphors with energy transfer. Proc IEEE. 1973;61(6):758–86.

Auzel F. Rare earth doped vitroceramics: new, efficient, blue and green emitting materials for infrared up-conversion. J Electrochem Soc. 1975;122(1):101.

Auzel F. Upconversion processes in coupled ion systems. J Lumin. 1990;45(1–6):341–345.

Auzel F. Upconversion and anti-Stokes processes with f and d ions in solids. Chem Rev. 2004;104(1):139–73.

Auzel F, Pecile D. Comparison and efficiency of materials for summation of photons assisted by energy transfer. J Lumin. 1973;8(1):32–43.

Auzel F, Pecile D. Absolute efficiency for IR to blue conversion materials and theoretical prediction for optimized matrices. J Lumin. 1976;11(5–6):321–30.

Bloembergen N. Solid state infrared quantum counters. Phys Rev Lett. 1959;2(3):84–5.

Chamarro MA, Cases R. Energy up-conversion in (Yb, Ho) and (Yb, Tm) doped fluorohafnate glasses. J Lumin. 1988;42(5):267–74.

Chen G, et al. Upconversion emission enhancement in Yb3+/Er3+-codoped Y2O3 nanocrystals by tridoping with Li+ ions. J Phys Chem C. 2008;112(31):12030–6.

Chivian JS, Case WE, Eden DD. The photon avalanche: a new phenomenon in Pr3+-based infrared quantum counters. Appl Phys Lett. 1979;35(2):124–5.

Cresswell PJ, Robbins DJ, Thomson AJ. Rhenium(IV) as a sensitizer for two-step blue up-converters. J Lumin. 1978;17(3):311–24.

Deng R, et al. Temporal full-colour tuning through non-steady-state upconversion. Nat Nanotechnol. 2015;10(3):237–42.

Feng W, Sun LD, Yan CH. Ag nanowires enhanced upconversion emission of NaYF4:Yb, Er nanocrystals via a direct assembly method. Chem Commun (Camb). 2009;29:4393–5.

Föster T. Intermolecular energy transfer and fluorescence. Ann. Phys. Leipzig. 1948;2:55–75.

Fujimoto Y, et al. Visible fiber lasers excited by GaN laser diodes. Prog Quantum Electron. 2013;37(4):185–214.

Garfield DJ, et al. Enrichment of molecular antenna triplets amplifies upconverting nanoparticle emission. Nat Photonics;2018.

Goldner P, Pellé F, Auzel F. Theoretical evaluation of cooperative luminescence rate in Yb3+: CsCdBr 3 and comparison with experiment. J Lumin. 1997;72–74:901–3.

Guo H, et al. Seed-mediated synthesis of NaY F4: Y b, Er/NaGdF4 nanocrystals with improved upconversion fluorescence and MR relaxivity. Nanotechnology. 2010;21(12):125602.

Guy S, Joubert MF, Jacquier B. Blue upconverted fluorescence via photon-avalanche pumping in YAG: Tm. physica status solidi (b). 1994;18(1).

Han S, et al. Enhancing luminescence in lanthanide-doped upconversion nanoparticles. Angew Chem Int Ed Engl. 2014;53(44):11702–15.

Hao J, Zhang Y, Wei X. Electric-induced enhancement and modulation of upconversion photoluminescence in epitaxial BaTiO3:Yb/Er thin films. Angew Chem. 2011;123(30):7008–12.

Heer S, et al. Blue, green, and red upconversion emission from lanthanide-doped LuPO4 and YbPO4 nanocrystals in a transparent colloidal solution. Angew Chem Int Ed Engl. 2003;42(27):3179–82.

Heer S, et al. Highly efficient multicolour upconversion emission in transparent colloids of lanthanide-doped NaYF4 nanocrystals. Adv Mater. 2004;16(23–24):2102–5.

Hehlen MP, et al. Encyclopedia of materials; science and technology, vol. 4. Elsevier Science Ltd;2001. p. 9956.

Hehlen MP, et al. Cooperative optical bistability in the dimer system Cs3Y2Br 9:10% Yb3+. J Chem Phys. 1996;104(4):1232–44.

Hehlen MP, et al. Site-Selective, Intrinsically Bistable Luminescence ofYb3 + Ion Pairs inCsCdBr 3. Phys Rev Lett. 1999;82(15):3050–3.

Huang SH. Luminescence kinetics of ion centers. Science Press;2002.

Huang X, et al. Enhancing solar cell efficiency: the search for luminescent materials as spectral converters. Chem Soc Rev. 2013;42(1):173–201.

Hubert S, et al. Up conversion process in U4+-doped ThBr 4 and ThCl4. J Solid State Chem. 1986;61(2):252–9.

Hutchinson JA, Allik TH. Diode array-pumped Er, Yb: phosphate glass laser. Appl Phys Lett. 1992;60(12):1424–6.

Joubert M-F. Photon avalanche upconversion in rare earth laser materials. Opt Mater. 1999;11(2–3):181–203.

Kaminskii AA. Laser crystals: their physics and properties[M], vol. 14. Berlin: Springer;2013.

Kaplyanskii AA, McFarlane RM. Spectro-scopy of solids containing rare earth ions. In: Modern problems in condensed matter sciences, vol. 21;1987.

Krasutsky NJ. 10-μm samarium based quantum counter. J Appl Phys. 1983;54(3):1261–7.

Kueny AW, Case WE, Koch ME. Nonlinear-optical absorption through photon avalanche. J Opt Soc Am B. 1989;6(4):639.

Kueny AW, Case WE, Koch ME. Infrared-to-ultraviolet photon-avalanche-pumped upconversion in Tm:LiYF_4. J Opt Soc Am B. 1993;10(10):1834.

Lee WI, Bae Y, Bard AJ. Strong blue photoluminescence and ECL from OH-terminated PAMAM dendrimers in the absence of gold nanoparticles. J Am Chem Soc. 2004;126(27):8358–9.

Li LL, et al. Biomimetic surface engineering of lanthanide-doped upconversion nanoparticles as versatile bioprobes. Angew Chem Int Ed Engl. 2012;51(25):6121–5.

Liu C, Chen D. Controlled synthesis of hexagon shaped lanthanide-doped LaF3 nanoplates with multicolor upconversion fluorescence. J Mater Chem. 2007;17(37):3875.

Liu C, et al. Monodisperse, size-tunable and highly efficient β-NaYF4:Yb, Er(Tm) up-conversion luminescent nanospheres: controllable synthesis and their surface modifications. J Mater Chem. 2009;19(21):3546.

Lüthi SR, et al. Excited-state dynamics and optical bistability in the dimer system Cs3Lu2Br 9: Yb3+. J Lumin. 1998;76–77:447–50.

Malinowski M, et al. Optical transitions of Ho3+ in YAG. J Alloy Compd. 2000;300–301:389–94.

Martín IR, et al. Infrared, blue and ultraviolet upconversion emissions in Yb3+–Tm3+-doped fluoroindate glasses. Spectrochim Acta Part A: Mol Biomol Spectrosc. 1999;55(5):941–5.

Méndez-Ramos J, et al. Optical properties of Er3+ ions in transparent glass ceramics. J Alloy Compd. 2001;323–324:753–8.

Moncorgé R, Auzel F, Breteau JM. Excited state absorption and energy transfer in the infrared laser material MgF2: Ni2+. Philos Mag B. 2006;51(5):489–99.

Nadort A, Zhao J, Goldys EM. Lanthanide upconversion luminescence at the nanoscale: fundamentals and optical properties. Nanoscale. 2016;8(27):13099–130.

Nakazawa E, Shionoya S. Cooperative Luminescence in YbPO4. Phys Rev Lett. 1970;25(25):1710–2.

O'Brien B. Development of infra-red sensitive phosphors. JOSA. 1946;36(7):369–71.

Ofelt GS. Intensities of crystal spectra of rare-earth ions. J Chem Phys. 1962;37(3):511–20.

Page RH, et al. Upconversion-pumped luminescence efficiency of rare-earth-doped hosts sensitized with trivalent ytterbium. In: Advanced solid state lasers, vol. 3;1998.

Park YI, et al. Nonblinking and nonbleaching upconverting nanoparticles as an optical imaging nanoprobe and T1 magnetic resonance imaging contrast agent. Adv Mater. 2009;21(44):4467–71.

Pelletier-Allard N, Pelletier R. An internal quantum counter for lifetime measurements. Opt Commun. 1991;81(3–4):247–50.

Qian HS, Zhang Y. Synthesis of hexagonal-phase core-shell NaYF4 nanocrystals with tunable upconversion fluorescence. Langmuir. 2008;24(21):12123–5.

Qin X, Yokomori T, Ju Y. Flame synthesis and characterization of rare-earth (Er3+, Ho3+, and Tm3+) doped upconversion nanophosphors. Appl Phys Lett. 2007;90(7):073104.

Rohwer LS, Martin JE. Measuring the absolute quantum efficiency of luminescent materials. J Lumin. 2005;115(3–4):77–90.

Salley GM, Valiente R, Guedel HU. Luminescence upconversion mechanisms in Yb3+–Tb3+ systems. J Lumin. 2001;94–95:305–9.

Sangeetha NM, van Veggel FCJM. Lanthanum Silicate and Lanthanum Zirconate Nanoparticles Co-Doped with Ho3+ and Yb3+: Matrix-Dependent Red and Green Upconversion Emissions. J Phys Chem C. 2009;113(33):14702–7.

Schäfer H, et al. Synthesis and optical properties of KYF4/Yb, Er nanocrystals, and their surface modification with undoped KYF4. Adv Func Mater. 2008;18(19):2913–8.

Singh AK, et al. Photon avalanche upconversion and pump power studies in LaF3:Er3+/Yb3+ phosphor. Appl Phys B. 2011;104(4):1035–41.

Song E, et al. Room-Temperature Wavelength-Tunable Single-Band Upconversion Luminescence from Yb3+/Mn2+ Codoped Fluoride Perovskites ABF3. Advanced Optical Materials. 2016;4(5):798–806.

Suyver JF, et al. Novel materials doped with trivalent lanthanides and transition metal ions showing near-infrared to visible photon upconversion. Opt Mater. 2005;27(6):1111–30.

Tang J, et al. Selectively enhanced red upconversion luminescence and phase/size manipulation via Fe(3+) doping in NaYF4:Yb. Er nanocrystals. Nanoscale. 2015;7(35):14752–9.

Vetrone F, et al. Significance of Yb3+ concentration on the upconversion mechanisms in codoped Y2O3:Er3+, Yb3+ nanocrystals. J Appl Phys. 2004;96(1):661–7.

Wang F, Liu X. Recent advances in the chemistry of lanthanide-doped upconversion nanocrystals. Chem Soc Rev. 2009;38(4):976–89.

Wang F, Wang J, Liu X. Direct evidence of a surface quenching effect on size-dependent luminescence of upconversion nanoparticles. Angew Chem Int Ed Engl. 2010;49(41):7456–60.

Wu X, et al. Infrared-to-visible conversion luminescence of Tm 3+ and Yb 3+ ions in glass ceramics. J Lumin. 1994;60–61:212–5.

Xu X, Su M. Luminescence and luminescient materials. Chemical Industry Press;2004.

Yen WM, Selzer PM. Laser Spectroscopy of solids. Springer;1981.

Yi GS, Chow GM. Synthesis of hexagonal-phase NaYF4:Yb, Er and NaYF4:Yb, Tm nanocrystals with efficient up-conversion fluorescence. Adv Func Mater. 2006;16(18):2324–9.

Yi G, Peng Y, Gao Z. Strong red-emitting near-infrared-to-visible upconversion fluorescent nanoparticles. Chem Mater. 2011;23(11):2729–34.

Zhao T, et al. Upconversion nanocrystals doped glass: a new paradigm for integrated optical glass;2016. p. AM5C.7.

Zhao C, et al. Li+ ion doping: an approach for improving the crystallinity and upconversion emissions of NaYF4:Yb3+, Tm3+ nanoparticles. Nanoscale. 2013a;5(17):8084–9.

Zhao J, et al. Single-nanocrystal sensitivity achieved by enhanced upconversion luminescence. Nat Nanotechnol. 2013b;8(10):729–34.

Zhou J, Liu Z, Li F. Upconversion nanophosphors for small-animal imaging. Chem Soc Rev. 2012;41(3):1323–49.

Zhu X, et al. Anti-stokes shift luminescent materials for bio-applications. Chem Soc Rev. 2017;46(4):1025–39.

Chapter 2
Synthesis and Preparation of Upconverting Phosphor Particles

Bo Qiao, Suling Zhao and Yan Zheng

Abstract In this chapter, the synthesis and preparation of upconverting phosphor particles will be introduced. The components of rare earth upconversion luminescence materials are the activator, the sensitizer, and the host. There are many kinds of upconversion luminescence materials that have been discovered so far. Although they are mainly related compounds of rare earth elements, they can be classified into vanadates, phosphates, sulfides, sulfur oxides, oxides, oxyhalides, molybdates, etc. At present, the synthesis of high-quality rare earth halide upconversion luminescence nanomaterials mainly includes precipitation method, water/solvothermal method, pyrolysis method and sol gel method. In order to apply UCNPs biologically, the hydrophilic modification of the surface, such as addition of carboxyl, amino or aldehyde groups, is very important and necessary. At the end, the main challenges of rare earth upconversion luminescence materials will be discussed.

Keywords Upconversion luminescence materials · Rare earth elements · Synthesis processes · Surface modification · Cytotoxicity

2.1 Introduction

The Upconversion luminescence is the phenomenon of absorbing two or more low-energy infrared photons and emitting high-energy visible or even ultraviolet photons by doping with different rare earth ions in host materials and excited with infrared

---

B. Qiao · S. Zhao (✉)
Key Laboratory of Luminescence and Optical Information, Beijing Jiaotong University, Ministry of Education, Beijing 100044, China
e-mail: slzhao@bjtu.edu.cn

B. Qiao
e-mail: boqiao@bjtu.edu.cn

Institute of Optoelectronics Technology, Beijing Jiaotong University, Beijing 100044, China

Y. Zheng
Shanghai Kerune Phosphor Technology Co., Ltd, Shanghai, China
e-mail: zy@kpt.net.cn

R. Yang (ed.), *Principles and Applications of Up-converting Phosphor Technology*,
https://doi.org/10.1007/978-981-32-9279-6_2

light to achieve red, green, blue and even the ultraviolet light emission. Due to the unique electronic structure of rare earth ions, rare earth ions have excellent fluorescence properties.

In the upconversion luminescence materials, the luminescent centers are mostly rare earth ions, such as Tm^{3+}, Er^{3+}, Ho^{3+}, Pr^{3+}, Nd^{3+}, and Sm^{3+}. Because rare earth ions have abundant energy levels, especially the crystal field of host materials makes each energy level further split, the increasing level density results in the increase of energy level matching opportunities. At the same time, the energy transfer between rare earth ions and the concentration quenching is more likely to occur. Therefore, it is necessary to consider the upconversion process to select the appropriate excitation wavelength and level matching as well as to select host materials with suitable phonon energy to avoid non-radiative transitions. In order to improve the absorption of infrared light, it is often co-doped with sensitizer ions in the host materials. The most common sensitizer is Yb^{3+} because the energy level structure of Yb^{3+} ions is simple and easy to match the energy levels of other rare earth ions.

There are hundreds of rare earth ion doped upconversion luminescence materials, according to the different material components, they can be divided into fluoride, oxyhalide, oxide, and composite oxide. Among them, fluoride plays an important role and the upconversion luminescence efficiency is obviously higher than that of other materials due to low phonon energy of fluoride host materials. According to the different material structures, host materials can also be divided into glass, ceramic, polycrystalline, and single crystal. According to the size of materials, they can be divided into bulk materials, micro-materials, and nanomaterials. With the rise of nanotechnology, the nanocrystallization and its related research has gradually become a hot spot. Compared with the traditional fluorescent or quantum dot materials, the rare earth doped upconversion luminescence nanomaterials use near infrared-light as the excitation light source, which is a special rare earth doped luminescence material and mainly characterized as a typical anti-Stokes luminescence material. Besides, it has a larger anti-Stokes shift, which is different from the common Stokes luminescent materials. The infrared light as the excitation source is different from the traditional ultraviolet light source, which inherits many excellent characteristics of the infrared light. Therefore, the application of the upconversion luminescent nanomaterials has unique advantages in the biological field. First, the infrared light in biological optical window can effectively penetrate biological tissue. Second, organic materials of biological tissues have very low absorption to the near-infrared light and are not easily excited, then they are less likely to generate upconversion luminescence. Therefore, the upconversion luminescence signal generated by the nano upconverting luminescence materials in the living body has extremely high Signal-to-noise ratio, a good probe for bioimaging and medical detection. Third, the ultraviolet light is more harmful to the biological tissue, while the near-infrared light is relatively safe and even higher power laser light irradiates the biological tissue only generates a lower heating effect. All of these advantages make the upconversion luminescent materials very broad prospects in biological applications such as bioimaging, biodetection, photodynamic therapy, and targeted transportation. In addition, the upconversion luminescent nanoparticles also have a wide range of applications in the fields of solar cells, 3D displays, ion detection, and illumination.

2.2 The Components of Rare Earth Upconversion Luminescence Materials

As early as the late 1950s, upconversion of CdS luminescence has been observed (Yi and Chow 2007). Auzel (Zhang et al. 2012) discovered the effect of sensitizer Yb^{3+} on the luminescent properties of many luminescent centers such as Ho^{3+}. Since then, people have begun to analyze and characterize various properties of upconversion luminescent materials both theoretically and experimentally. Especially since the report of the upconversion nanomaterials by Professor Gudel in 2004 (Heer et al. 2004), the study of upconversion luminescence is re-up boom. Whether bulk materials or nanomaterials, upconversion luminescence materials mainly consist of two major components: the host material and the luminescence center, where the luminescence center includes the sensitizer and the activator, and only the luminescence center with a longer excited state lifetime can generate effective upconversion luminescence. Lanthanide rare earth ions have a longer excited state lifetime than that of common ions, so lanthanide rare earth ions are usually selected as luminescence centers to obtain high-efficiency and stable upconversion luminescence by through mono-doping or double ions co-doping. Mono-doped up-conversion luminescent materials are usually composed of the activator and the host, and co-doped is based on the activator, the sensitizer, and the host. The following will introduce each of these three components:

2.2.1 *The Host*

The host is the main component of the upconversion luminescence materials. Normally, the host does not contribute to the luminescence, that is, does not participate in the transition between the electron levels of the luminescence center, its main role is to provide a suitable crystal field environment around the luminescent center to produce a specific emission. However, the crystal structure of the host material determines the distance between the doping ions and their spatial positions, and also determines the coordination of the doping ions and the type of anions around them. Therefore, the interaction between the host lattice and dopant ions influences the upconversion luminescence properties of the dopant ions. In general, the choice of host materials based on the following aspects: 1. The host material for the excitation and emission of light must have high transparency. Therefore, most of the host materials have a wide bandgap that allows transmission of ultraviolet to infrared light. For example, YF_3 can provide a wide bandgap of >10 eV. 2. The host material can be doped with lanthanide rare earth ions, and the solubility of lanthanide rare earth ions in the host should be high. 3. The low phonon energy of the host material is required to minimize deleterious non-radiative relaxations that weaken the upconversion luminescence efficiency; 4. The host material needs to have an excellent chemical stability and thermal stability to retain the original crystal structures. 5. The host material needs to have a certain mechanical strength to reduce the possibility of mechanical damage. All trivalent rare earth ions show similar ionic radii and chemical properties. Therefore, their inorganic compounds are ideal host materials for lanthanide-based rare earth ions doping to achieve upconversion luminescence. Y, Gd, Lu, and La are usually selected as host elements. In addition, alkaline earth metal ions such as (Ca^{2+}, Sr^{2+}, Ba^{2+}) and some transitional metal ions such as (Zr^{4+}, Ti^{4+}) also have similar ionic radius as lanthanide ions (Wang and Liu 2009). Therefore, some inorganic compounds containing these ions, which can avoid crystal defects and lattice stress when they are doped with lanthanide ions, are also frequently used as upconversion luminescent host materials (Wang et al. 2009a; Zheng et al. 2013a; Patra et al. 2002, 2003) such as $NaYF_4$ (Baldini et al. 2007), Y_2O_3 (Vetrone et al. 2004), YF_3, CaF_2 (Wang et al. 2009a). However, lanthanide rare earth ion doping in these materials is usually accompanied by defects such as cationic vacancy defects, while interstitial anions are often used to balance the charge neutrality.

It has been demonstrated in Sect. 2.1 that the transitions between 4f-4f levels of the trivalent lanthanide rare earth ions (Ln^{3+} ions) are parity forbidden. When they are doped in compound and surrounded by inorganic lattices, the effect of crystal field to them is different due to that are in different positions in the crystal lattice. In the asymmetrical crystal field, the f-state of Ln^{3+} ions can interact with other ion states with opposite parity, resulting in the mixture configuration of odd-parity. Then the parity forbidden can be broken partially. This mixture configuration comes from the point-symmetry change and is favorable for 4f-4f energy level transitions of Ln^{3+} ions, which makes the originally forbidden 4f-4f electric dipole transitions become permissible, so the crystal structure of the host material has a great influence on the upconversion efficiency. The improvement of the asymmetry of the host crystal is an important means to improve the upconversion luminescence efficiency. In the low-symmetry rare earth ion-doped host material, such as in the hexagonal β-phase $NaYF_4$, the absorption and emission cross-sections of rare earth ions are about an order of magnitude larger than that in the inversion-symmetrical α-phase $NaYF_4$, so that the upconversion light emission intensity is strong in β-phase $NaYF_4$ (Krämer et al. 2004). Therefore, the upconversion luminescence can be regulated by adjusting the local crystal field of luminescent centers in inorganic host materials, but this adjustment may change the spatial distance between the luminescent centers, resulting in excessive multiphonon cross-relaxation and other energy transfer process. Another very effective way is to dope other optically inactive ions to compensate the side effects of crystal field regulation. In general, the main consideration in selecting these inactive ions is whether the cation radius matches the valence state. Many alkali metals and transition metal ions are often used to regulate the local crystal field of rare earth ions in the host. Because Li^+ ions have a relatively small cation radius, they are considered to be randomly distributed on the lattice sites or in the lattice as interstitial ions, which makes Li^+ suitable for adjusting the local crystal field of the host lattice. Zhang et al. first reported the doping of Li^+ ions in Y_2O_3 to increase the upconversion luminescence of Er^{3+} by 25 times. This is mainly because Li^+ doping breaks the symmetry of the crystal field around the rare earth ions (Chen et al. 2008). Subsequently, various studies have been reported to enhance upconversion luminescence by adjusting the local crystal field of rare earth ions in different matrices. For example, $NaYF_4$:Yb,Er nanoparticles are doped with 80% Li^+ ions, and the upconversion emission is enhanced 30 times (Wang and Nann 2009); By assuring that the particle size does not change substantially, the upconversion luminescence intensity is increased by co-doping Li^+ and K^+ in $NaYF_4$:Yb, Er (Liang et al. 2017).

In addition, the crystal structure of the host determines the phonon diagram of the material, which describes the allowed vibration modes in the crystal. Excited electrons can be relaxed to low energy levels by emitting phonons or phonon-assisted nonresonant energy transfer. Fluorides are often used as host materials because they have relatively low phonon energy ($\leq$600 cm^{-1}), which limits the path of multi-phonon nonradiative relaxation.

2.2.2 *The Activator*

The activator is the luminescent center of the upconversion luminescent nanomaterial. As the source of fluorescence in the whole material, it is required that the activator ion has rich energy levels, and its intermediate state needs a longer lifetime. When electrons in the ground state of activator ions absorb energy and are excited to the intermediate state, these electrons in the intermediate state can jump further to a higher level by absorbing the excitation photons or by energy transfer from excited sensitizers or other excited activator ions. Then de-excitation back to the ground state and upconversion light emission occurs. In addition, in order to avoid the quenching due to the cross-relaxation between activator ions, the doping concentration of the activator should not be too high, that is, the average distance between the activator ions should not be too close.

Trivalent lanthanide rare earth ions are the best activators for upconversion luminescence. Their absorption and emission spectra are mainly due to transitions of the 4f electrons, and the interaction between the 4f electrons and the host is small. Because these rare earth ions all have abundant, step-like energy levels, the energy difference between different energy levels corresponds to the energy of infrared photons or visible photons, respectively. The splitting of the same energy level is very narrow, at about 10–100 cm^{-1}. In addition, intermediate energy levels of many lanthanide rare earth ions have long lifetime of about 10 μs–10 ms, which favors electrons at these levels to jump further to higher energy levels by absorbing energy and emit light. The major emissions and corresponding energy transitions for making multi-colored UCNPs by using different dopant ions are listed in Table 2.1. Lanthanide

Table 2.1 The major emissions and corresponding energy transitions for making multicolored UCNPs by using different dopant ions (Heer et al. 2004; Naccache et al. 2008; Maciel et al. 2009; Pelle et al. 2011; Joshi and Rai 2012; Ming et al. 2013; Dey et al. 2012; Ramakrishna et al. 2013; Wang et al. 2011a; Su et al. 2012; Wang et al. 2013; Cao et al. 2008; Song et al. 2014; Liu and Chen 2007; Qin et al. 2007; Ehlert et al. 2008; Yi and Chow 2006; Mahalingam et al. 2009; Zhang et al. 2011)

Activators	Major emissions (nm)	Energy transitions	References
Pr^{3+}	489, 526, 548, 618, 652, 670, 732, 860	${}^3P_0 \rightarrow {}^3H_4$, ${}^1I_6 \rightarrow {}^3H_5$, ${}^3P_0 \rightarrow {}^3H_5$, ${}^3P_0 \rightarrow {}^3H_6$, ${}^3P_0 \rightarrow {}^3F_2$, ${}^3P_1 \rightarrow {}^3F_3$, ${}^3P_0 \rightarrow {}^3F_4$, ${}^1I_6 \rightarrow {}^1G_4$	Naccache et al. (2008), Maciel et al. (2009), Pelle et al. (2011), Joshi and Rai (2012), Ming et al. (2013)
Nd^{3+}	430, 482, 525, 535, 580, 600, 664, 766	${}^2P_{1/2} \rightarrow {}^4I_{9/2}$, ${}^2P_{1/2} \rightarrow {}^4I_{11/2}$, ${}^2P_{1/2} \rightarrow {}^4I_{13/2}$, ${}^4G_{7/2} \rightarrow {}^4I_{9/2}$, ${}^2P_{1/2} \rightarrow {}^4I_{15/2}{}^4G_{7/2} \rightarrow {}^4I_{11/2}$, ${}^2G_{7/2} \rightarrow {}^4I_{9/2}$, ${}^4G_{7/2} \rightarrow {}^4I_{13/2}$, ${}^2G_{1/2} \rightarrow {}^4I_{15/2}$	Dey et al. (2012), Ramakrishna et al. (2013)
Sm^{3+}	555, 590	${}^4G_{5/2} \rightarrow {}^6H_{5/2}$, ${}^4G_{5/2} \rightarrow {}^6H_{7/2}$	Wang et al. (2011a), Su et al. (2012)
Eu^{3+}	590, 615, 690	${}^5D_0 \rightarrow {}^7F_1$, ${}^5D_0 \rightarrow {}^7F_2$, ${}^5D_0 \rightarrow {}^7F_4$	Wang et al. (2011a), Su et al. (2012)
Gd^{3+}	204, 254, 278, 306, 312	${}^6G_{7/2} \rightarrow {}^8S_{7/2}$, ${}^6G_{7/2} \rightarrow {}^8S_{7/2}$, ${}^6I_J \rightarrow {}^8S_{7/2}$, ${}^6P_{5/2} \rightarrow {}^8S_{7/2}$, ${}^6P_{7/2} \rightarrow {}^8S_{7/2}$	Wang et al. (2013), Cao et al. (2008), Song et al. (2014)
Tb^{3+}	490, 540, 580, 615	${}^5D_4 \rightarrow {}^7F_6$, ${}^5D_4 \rightarrow {}^7F_5$, ${}^5D_4 \rightarrow {}^7F_4$, ${}^5D_4 \rightarrow {}^7F_3$	Wang et al. (2011a), Su et al. (2012)
Dy^{3+}	570	${}^4F_{9/2} \rightarrow {}^4H_{13/2}$	Wang et al. (2011a), Su et al. (2012)
Ho^{3+}	542, 645, 658	${}^5S_2 \rightarrow {}^5I_8$, ${}^5F_5 \rightarrow {}^5I_8$	Liu and Chen (2007), Qin et al. (2007), Ehlert et al. (2008)
Er^{3+}	411, 523, 542, 656	${}^2H_{9/2} \rightarrow {}^2I_{15/2}$, ${}^2H_{11/2} \rightarrow {}^2I_{15/2}$, ${}^2S_{3/2} \rightarrow {}^4I_{15/2}$, ${}^2F_{9/2} \rightarrow {}^4I_{15/2}$	Liu and Chen (2007), Yi and Chow (2006)
Tm^{3+}	294, 345, 368, 450, 475, 650, 700, 800	${}^1I_6 \rightarrow {}^3H_6$, ${}^1I_6 \rightarrow {}^3F_4$, ${}^1D_2 \rightarrow {}^3H_6$, ${}^1D_2 \rightarrow {}^3F_4$, ${}^1G_4 \rightarrow {}^3H_6$, ${}^1G_4 \rightarrow {}^3F_4$, ${}^3F_3 \rightarrow {}^3H_6$, ${}^3H_4 \rightarrow {}^3H_6$	Mahalingam et al. (2009), Zhang et al. (2011)

rare earth ions such as Pr^{3+}, Tb^{3+}, Sm^{3+}, Nd^{3+}, Ho^{3+}, Tm^{3+}, and Er^{3+} are common activator ions. Among these rare earth ions, Er^{3+} has the best upconversion luminescence properties and is studied widely. In addition to $La^{3+}(4f^0)$, $Ce^{3+}(4f^1)$ and $Lu^{3+}(4f^{14})$ ions, other lanthanide rare earth ions, can also generate upconversion luminescence through appropriate energy transfer processes. However, the upconversion luminescence properties of some rare earth ions, such as Nd^{3+}, Pr^{3+}, and Sm^{3+}, are weak because that there is a large probability of non-radiative relaxation inside these ions. Due to the lack of long-lived intermediate states, $Eu^{3+}$and Tb^{3+} do not absorb the appropriate energy to allow electrons to jump to higher energy levels, and thus cannot be upconverted by energy transfer processes. But upconversion luminescence was also observed in Yb^{3+}–Eu^{3+}/Yb^{3+}–Tb^{3+} systems, of which the mechanism was explained as a CET process (Martín-Rodríguez et al. 2009). Because Yb^{3+} lacks a step-like excited state energy level, although it is often used as a sensitizer, upconversion luminescence from the Yb^{3+}–Yb^{3+} complex has also been observed in single-doped Yb^{3+} nanomaterials (Nakazawa 1976). Another kind of special ion is Gd^{3+} ion, its first excited state energy (about 32224 cm^{-1}) is much higher than other Ln^{3+} ions, so it is difficult to achieve upconversion luminescence through energy transfer between other ions, even if the Yb^{3+}–Yb^{3+} complex also transfer their energy difficultly to Gd^{3+}. The upconversion luminescence observed on Gd^{3+} reported by the literature is derived from the higher excited state energy transfer of Tm^{3+} (Dong et al. 2013).

Due to the complex and dense 4f levels, the doping concentration of activator rare earth ions cannot be too high. The low concentration doping can avoid unnecessary multiphonon assisted cross-relaxation. However, the upconversion luminescence nanomaterials with single rare earth ion doped do not have high absorption efficiency for infrared light for excitation, which therefore have low up-conversion luminescence efficiency. The excessively high doping concentration results in fluorescence quenching. Therefore, it is necessary to add a sensitizer to enhance the absorption of infrared light to enhance the up-conversion luminescence efficiency.

2.2.3 *The Sensitizer*

In the single ion-doped upconversion luminescent materials, the concentration of dopant ion determines the spacing between ions and the absorption of excitation light. In the case of high doping concentration, more excitation light can be absorbed, but at the same time the cross-relaxation process is enhanced and results in luminescence quenching; at lower doping concentrations, the absorption of light is relatively limited, and the absorption cross-section of many lanthanide metal ions is low, which is not conducive to the absorption of excitation light. Therefore, in order to improve the luminescence efficiency, an ion having a large light absorption cross-section is often used as a sensitizer co-doped with a luminescent center ion, and the upconversion luminescence is realized by energy transfer from the sensitizer to the luminescence center (activator).

Because sensitizer ions can absorb excitation photons more efficiently and transfer their own energy to activator ions, this requires that the sensitizer must have a large absorption cross-section to the infrared light and a suitable energy level to match the energy level of the activator ions. Among all the Ln^{3+} ions, the energy level structure of Yb^{3+} is relatively simple. There are only two energy levels $^2F_{7/2}$–$^2F_{5/2}$, of which the energy difference corresponds to the energy of 980 nm photons. The absorption cross-section of Yb^{3+} is the biggest among all Ln^{3+} ions, about 9.11×10^{-21} cm^{-2}. Hence, it is a good sensitizer ion, and often used as the sensitizer in current upconversion luminescence nanomaterial systems.

The rare earth ion Yb^{3+} with a simple energy level structure can tolerate high doping concentration and thus absorb more excitation photons. In contrast, the doping concentration of activator ions is relatively low, such as less than 2 mol%, to reduce the loss energy resulted from the inter-ion cross-relaxation processes. After the sensitizer absorbs the excitation photons, it can effectively transfer the obtained energy to the activator ion, such as Er^{3+}, Tm^{3+}, thereby improving the upconversion luminescence efficiency. It was found that Yb^{3+}–Ho^{3+}, Yb^{3+}–Tm^{3+} and Yb^{3+}–Er^{3+} are common co-doped systems. In these co-doped systems, there is also an optimal doping concentration of Yb^{3+} at 20 mol%. If the concentration is too high, the energy transfer between Yb^{3+}–Yb^{3+} ions will occur and eventually lead to reduced upconversion luminescence.

2.3 Major Types of Rare Earth Upconversion Luminescence Nanomaterials

Since the upconversion luminescence has been reported, the infrared to visible upconversion luminescence of different rare earth ions doped crystals, glass, and ceramic host materials has been widely studied. For host materials, not only is the material required to have a low phonon energy, good optical properties, but it is also required a certain mechanical strength and chemical stability. Therefore, finding suitable host materials is very important for obtaining stable and efficient upconversion luminescence. There are many kinds of upconversion luminescence materials that have been discovered so far. Although they are mainly related compounds of rare earth elements, they can be classified into vanadates, phosphates, sulfides, sulfur oxides, oxides, oxyhalides, molybdates, etc., some of which have been nano-sized. In many nanomaterials, sulfides, bromides, and chlorides have low phonon energy, but because of the harsh reaction synthesis conditions or the special nature of materials, their large-scale applications are limited. Oxides cannot realize efficient upconversion luminescence due to their high phonon energy, so the use of oxides has encountered obstacles. Fluorides have become the focus of attention due to their high stability, low phonon energy, and being easily doped by rare earth ions.

A very important parameter for upconversion luminescence nanomaterials in the biological field is the particle size. Typical lanthanide-doped hosts with different sizes, crystal structures, and morphologies are listed in Table 2.2. In general, the crystal structure and particle size relate to the parameters of the host nanomaterial. The current reported various doped rare earth ion upconversion luminescence nano-

Table 2.2 Typical lanthanide-doped hosts with different sizes, crystal structures and morphologies (Yi and Chow 2006; Mahalingam et al. 2009; Chen et al. 2010; Ostrowski et al. 2012; Liu et al. 2011; Mai et al. 2006; Li et al. 2008; Park et al. 2009; Wang et al. 2010a; Johnson et al. 2011; Sarkar et al. 2013; Wong et al. 2011; Wang et al. 2014; Huang et al. 2014; Wang et al. 2011b; Yi et al. 2011; Zheng et al. 2013b; Chen et al. 2012a)

Host:dopant	Size (nm)	Crystal structure	Morphology	References
$YLiF_4$:Yb,Tm	~80	Tetragonal	Diamond	Mahalingam et al. (2009)
$NaYF_4$:Yb,Er	13.6	Cubic	Polyhedron	Mai et al. (2006)
$NaYF_4$:Yb,Er	187 × 71	Hexagonal	Hexagonal plates	Mai et al. (2006)
$NaYF_4$:Yb,Er	21 ± 0.5	Hexagonal	Spheres	Li et al. (2008)
$NaYF_4$:Yb,Er	10.5 ± 0.7	Hexagonal	Spheres	Yi and Chow (2006)
$NaYF_4$:Yb,Er	4.5–15	Hexagonal	Spheres	Ostrowski et al. (2012)
$NaYF_4$:Yb,Tm	~7	Cubic	Spheres	Chen et al. (2010)
$NaGdF_4$:Yb,Er	20, 31, 4	Cubic	Cube	Park et al. (2009)
$NaGdF_4$:Yb,Tm	10, 15, 2	Hexagonal	Spheres	Wang et al. (2010a)
$NaGdF_4$	2.5–8	Hexagonal	Hexagonal	Johnson et al. (2011)
$NaLuF_4$:Yb,Er	18.9	Cubic	Spheres	Liu et al. (2011)
$NaLuF_4$:Gd,Yb, Tm	7.81–17	Hexagonal	Spheres	Liu et al. (2011)
$BaLuF_5$	<5	Cubic	Dot	Sarkar et al. (2013)
$KGdF_4$:Yb,Tm	~3.7	Cubic	Irregular	Wong et al. (2011)
KYb_2F_7:Er	109 × 26 × 12	Orthorhombic	Ice lolly stick	Wang et al. (2014)
$LiLuF_4$:Yb,Er	28 ± 1.5	Tetragonal	Rhomboid	Huang et al. (2014)
$KmnF_3$:Yb,Er	~30	Cubic	Cube	Wang et al. (2011b)
YOF:Yb,Er	15 ± 0.4	Cubic	Spheres	Yi et al. (2011)
CaF_2:Yb,Er	3.8 ± 0.5	Cubic	Cube	Zheng et al. (2013b)
YF_3:Yb,Er	~3.7	Orthorhombic	Irregular	Chen et al. (2012a)

materials are all between 20 and 50 nm in size. However, for in vivo biodetection, the ideal nanoparticle size is less than 10 nm, so that nanoparticles can be effectively removed from the body. Therefore, how to obtain upconverting luminescence particles having a size less than 10 nm and having a strong upconversion luminescence property is very critical for further promoting upconversion luminescence for in vivo detection. In response to this problem, a series of results have been achieved. For example, Prasad group reported the synthesis of α-$NaYF_4$ upconversion nanomaterials with a monodisperse size between 7 and 10 nm, which obtained a high upconversion luminescence (Chen et al. 2010). As the concentration of Yb^{3+} ions rise from 20 to 100%, the near-infrared (NIR) emission intensity can be increased by 43 times. This $NaYbF_4$:Tm^{3+} nanomaterial with such small particle size has 3.6 times stronger upconversion luminescence intensity than $NaYbF_4$:Tm^{3+} with a particle size of 25–30 nm. Cohen's group obtained β-$NaYF_4$:Yb,Er nanoparticles with a size of 4.5–15 nm by controlling the synthesis process. These particles with a core-shell structure of β-$NaYF_4$:Yb, Er@$NaYF_4$ less than 10 nm show an upconversion luminescence efficiency of (0.18 ± 0.01%) that is higher than that of 37 nm β-$NaYF_4$:Yb,Er at (0.14 ± 0.01%) (Ostrowski et al. 2012). Recently, Liu et al. synthesized $NaLuF_4$ upconversion nanoparticles with a hexagonal phase of less than 10 nm, and obtained an upconversion luminescence with a quantum efficiency of 0.47 ± 0.06% (Liu et al. 2011).

The upconversion luminescence properties is another crucial parameter for upconversion luminescence nanomaterials. As the particle size decreases, the upconversion luminescence drops sharply because of the large proportion of doped rare earth ions distributed on the surface of the nanoparticles. The luminescence quenches due to surface defects and high energy oscillations caused by weak binding of impurities, ligands, and solvents on the surface of the particles. In addition, rare earth ions that are excited inside the particles can also transfer energy to nearby surface rare earth ions, causing non-radiation transitions. In order to avoid this phenomenon, a layer of host-lattice-matched shells on the surface of nanoparticles is a method for effectively improving upconversion luminescence. This method can reduce the non-radiative relaxation of ions on the surface. In this core-shell structure, the doped rare earth ions are confined in the core, and the shell layer can effectively suppress the energy loss on the particle surface to enhance the upconversion luminescence performance. This method was first reported by Lezhnina and his collaborators, and the core-shell upconversion nanostructures were realized on EuF_3@GdF_3, GdF_3@EuF_3, LaF_3:Yb, Ho@LaF_3, LaF_3:Nd@LaF_3 (Lezhnina et al. 2006). In 2007, Chow et al. achieved approximately 7–29 times enhancement of upconversion luminescence by preparing a layer of $NaYF_4$ on the surface of $NaYF_4$:Yb,Er and $NaYF_4$:Yb,Tm nanocrystals. Since then, many researchers have prepared different types of upconversion luminescence core-shell nanomaterials, such as $NaYF_4$:Yb, Er(Tm)@$NaGdF_4$ (Zhang et al. 2012; Guo et al. 2010), $NaYF_4$:Yb,Er(Tm)@$NaYF_4$ (Mai et al. 2007; Wang et al. 2009b), $NaGdF_4$:Yb, Tm(Er,Nd)@$NaGdF_4$ (Park et al. 2009; Wang et al. 2010b; Chen et al. 2012b). For the selection of the shell material, generally choose the composition, crystal structure or lattice as same as the core host material as the shell material, that is, the shell material is the same host material but without doping rare earth ions, which is a very effective design and method to enhance upconversion luminescence. There also are reports of rare earth ions being introduced into the shell. For example, Capobianco and his collaborators prepared $NaGdF_4$:Yb, Er@$NaGdF_4$:Yb core-shell nanoparticles (Vetrone et al. 2009), which not only achieved energy transfer from Yb to Er in the core., the Yb of the shell can also transfer energy to Er in the core. Then the stronger green and red upconversion luminescence of Er than 3 times and 10 times.

The upconverting luminescent nanomaterials are classified into the following five categories depending on the composition of the host materials.

2.3.1 Rare Earth Fluorides

Rare earth ion doped fluoride crystals and glasses (including optical fibers) are the focus and hotspot of upconversion luminescence research. This is mainly due to the low phonon energy of fluoride hosts, especially the low frequcncy of phonons in heavy metal fluoride hosts. Hence, the non-radiative transition probability of the rare earth ion and the non-radiative cross-relaxation probability between ions are reduced, which is favor to enhance the radiative transition probability and therefore the upconversion luminescence efficiency is high. Many fluorides, such as $NaYF_4$, LaF_3, YF_3, and CaF_2, can be used as upconversion luminescence host materials. Among them, $NaREF_4$ type fluorides are the most efficient upconversion luminescence host materials found so far. They have low phonon vibration energy (<400 cm^{-1}), low non-radiative relaxation rate and high radiation emissivity. Among this kind fluoride hosts, the most studied is $NaYF_4$. There are two crystal phases in $NaYF_4$: cubic phase (α) and hexagonal phase (β). The upconversion luminescence efficiency of the hexagonal phase $NaYF_4$ is much higher than that of cubic phase $NaYF_4$. In this host material, doped rare earth ions such as Tm^{3+}, Ho^{3+}, Er^{3+}, etc. are used as activator ions, and their upconversion luminescence can be achieved under the excitation of infrared light. In 2006, Wang et al. prepared rare earth ion doped $NaYF_4$ nanoparticles (Patra et al. 2002). In 2006, Boyer et al. reported a new simple method for the synthesis of nano-$NaYF_4$:Er,Yb with a particle size of 10–50 nm (Patra et al. 2003). Wang et al. studied the effect of Gd^{3+} doping concentration on the crystal phase and luminescence properties of $NaYF_4$, and also adjusted the upconversion luminescence efficiency by changing the doping concentration of Gd^{3+} (Krämer et al. 2004). In 2009, Bogdan et al. used the thermal decomposition method to prepare $NaYF_4$:Er,Yb, and achieved the coating and dispersion of nanomaterials by removing the oleate ligands. Red and green emission of Er ions were obtained (Chen et al. 2008). In order to improve the upconversion luminescence intensity, Vetrone et al. prepared the $NaGdF_4$:Er, Yb@$NaGdF_4$ with a core-shell structure, improved the upconversion luminescence through the core-shell structure and realized the luminescence color regulation (Wang and Nann 2009; Dong et al. 2015).

It is well known that the method usually used to adjust the upconversion luminescence spectrum is by controlling the doping concentration or changing the type of doping ions. However, at high doping concentrations, the concentration quenching phenomenon in the excited states limits the upconversion luminescence properties. Therefore, low concentration doping of the rare earth ions must be ensured to reduce the concentration quenching. Liu et al. prepared a new series of KYb_2F_7 nanocrystals with an orthorhombic structure in which doped rare earth ions are distributed in a four-component array. This special distribution saves excitation energy in the sublattice region of the host, effectively reducing the transfer of excitation energy to defects. This shows that KYb_2F_7 not only can be used as a host material, but also can act as a sensitizer to absorb and preserve excitation energy (Wang et al. 2014). This result provides a good way for further improving the upconversion luminescence properties through the energy clusters in the host sublattice.

$NaGdF_4$:Yb,Er, as an upconversion nanomaterial, not only has good upconversion properties, excellent light stability and chemical stability, long lifetime luminescence, narrow emission band, low background fluorescence, and other common advantages, but also is an ideal T1 contrast agent that can be applied to both fluorescence imaging and magnetic resonance imaging due to the paramagnetic and a large amount of unpaired electrons from Gd^{3+} ions. Therefore, the upconversion nanoparticles $NaGdF_4$:Yb, Er also contain the advantages of high sensitivity, high resolution, low toxicity, real-time monitoring, etc., which have been the interest of the research (Chen et al. 2011).

Although the rare earth ion doped fluoride system has a high upconversion efficiency due to its low phonon energy, the disadvantages are also very obvious. The chemical stability is poor, the mechanical strength is low, the preparation environment is demanding high, and the production process is difficult and high cost. These shortcomings have limited its application to a certain extent.

2.3.2 Rare Earth Oxide Series

The rare earth oxide upconversion material has a high phonon energy and therefore has a low upconversion efficiency. However, it has the advantages of simple preparation process, low environmental conditions, and large range of components for forming glass phase, high solubility of rare earth ions, good mechanical strength, and good chemical stability. The typical rare earth oxide upconversion material is $Nd_2(W0_4)_3$, which converts 808 nm laser light to visible light at 457 and 657 nm at room temperature. Er^{3+} doped $YV0_4$ converts 808 nm laser light to 550 nm. Eu^{3+}, Yb^{3+} co-doped multicomponent silicate glass prepared by the sol-gel method can convert 973 nm light to orange light. Zhang et al. reported that the Y_2O_3:Yb,Er nanoparticle increases the upconversion luminescence intensity by doping with Li^+ ions. This is mainly due to the fact that Li^+ has a smaller ionic radius and facilitates doping. It is more favorable for breaking the symmetry of the crystal field around the rare earth ions by doping Li^+ ions and leads to PL enhancement (Chen et al. 2008).

2.3.3 Rare Earth Fluoride Oxide Series

As a kind of upconversion materials, the phonon energy of fluoride is small and the upconversion efficiency is high, but its biggest drawback is poor mechanical strength and chemical stability, which brings great difficulties to practical applications. In many host materials, the oxide host has good mechanical strength and chemical stability, but the phonon energy is large, and the efficiency of upconversion is low. Oxyfluoride synthesizes above two advantages, and thus has attracted great research attention. Compared with fluoride glasses, oxyfluoride glasses have much better laser loss threshold, chemical stability, and mechanical strength. Typical Er^{3+}-doped oxyfluoride glasses ($A1_20_3$, CdF_2, PbF_2, YF_3:Er^{3+}) have an excitation wavelength of 975 nm and an upconversion emission wavelength of 545, 660, and 800 nm.

2.3.4 Rare Earth Halide Series

This rare earth halide upconversion material mainly refers to rare earth ion doped heavy metal halides. Since they have lower vibrational energy, reduce the effect of multiphonon relaxation and improve the upconversion efficiency. However, because chlorides are prone to deliquescence in air, chloride glasses are extremely sensitive to moisture in the air, making it impossible to prepare and measure spectra in air.

2.3.5 Rare Earth Sulfide Series

Semiconductor nanomaterials such as ZnS can also be used as the upconversion host material. But in such nanomaterials, due to the mismatch of ionic radii, it has been debated whether the doped lanthanide rare earth ions mainly exist on the surface layer of the nanomaterials (Bol et al. 2002). Like rare earth fluorides, rare earth sulfide upconversion materials also have low phonon energy. However, it cannot be contacted with oxygen and water during the preparation and must be carried out in a closed condition. La_2S_3:Yb^{3+}, Pr^{3+} glass can upconvert 1064 nm excitation light to visible light in the 480–680 nm region at room temperature. Among them, Pr^{3+} is an upconversion ion and Yb^{3+} is a sensitizer.

In addition, the rare earth upconversion luminescence material also includes rare earth ion doped phosphate amorphous material system, fluoroborate glass material system, and germanate glass system.

In terms of upconversion luminescence efficiency, it is generally considered that chlorides>fluorides>oxides, which is simply considered from the phonon energy of the material, but the structure stability of the material is found to be chlorides<fluorides<oxides. This upconversion efficiency order is just the opposite of the order of structure stability of the material. Therefore, scientists have carried out

a series of studies and hope to find new host materials that have both the high upconversion efficiency as chlorides and fluorides and the structure stability as oxides, to achieve practical purposes.

In addition, the phase structure of the host material has a great influence on the upconversion luminescence efficiency. For example, the upconversion luminescence efficiency of the hexagonal phase $NaYF_4$:Yb,Er is one order of magnitude higher than that of its cubic phase material. This is because the phase structure of the host material is different, and the crystal field symmetry of the rare earth ions is also different. The host material of the low-symmetry phase structure has more unpaired components around the doped ions than the host material of the high-symmetry phase structure. This makes the doping ions have a higher transition probability.

2.4 Synthesis of Rare Earth Upconversion Luminescence Nanomaterials

At present, the synthesis of high-quality rare earth halide upconversion luminescence nanomaterials mainly includes precipitation method, water/solvothermal method, pyrolysis method, and sol-gel method as shown in Fig. 2.1. In order to prepare nanoparticles having a uniform particle size, controllable morphology, and high upconversion luminescence efficiency, it is sometimes necessary to combine the advantages of various methods. The raw materials for synthesizing rare earth upconversion luminescence nanoparticles are divided into precursors and stabilizers. Precursors are the core part of the nanoparticles, while stabilizers (also known as lig-

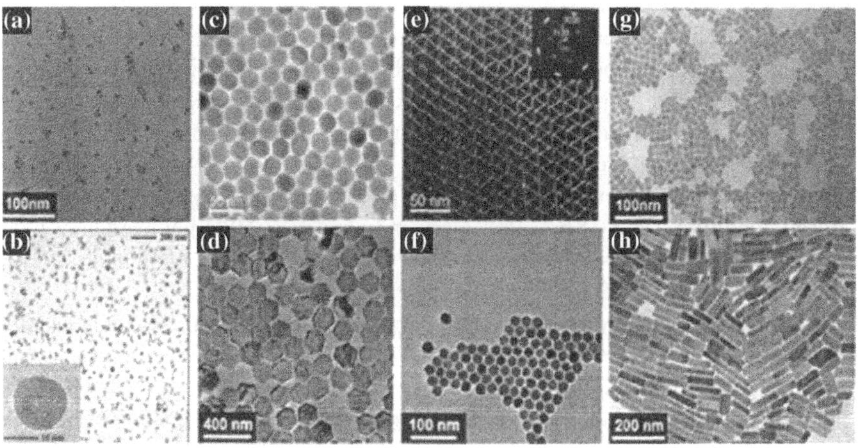

Fig. 2.1 TEM images of UCNPs. **a** LaF_3:Yb,Er, (solvothermal method), **b** $NaYF_4$:Yb,Er(Tm) (solvothermal method), **c** $NaYF_4$:Yb,Er(Tm)(precipitation method), **d** $NaYF_4$(thermal decomposition), **e** $NaYF_4$:Yb,Er (thermal decomposition), and **g**, **h** $NaYF_4$:Yb,Er (hydrothermal method)

ands) are used to prevent agglomeration of nanoparticles, regulate the particle size of nanoparticles, and protect the surface and slow down the continued growth of nanoparticles. The following is a brief introduction to the currently widely used methods.

2.4.1 The Precipitation Method

A precipitant is added to a soluble salt solution containing one or more ions to react to produce a sparingly soluble product that is precipitated from the solution. The excess ions in the original solution are washed off and dried to obtain the desired nanomaterial. This method is called as the precipitation method. If a precipitant is added to the above solution, all the ions are completely precipitated, which is called co-precipitation. The precipitation method is one of the most traditional methods, and rare earth nanoparticles with small particle size and narrow size distribution can be prepared. Compared with other methods, the precipitation method does not require complicated and expensive equipment, and the reaction conditions are mild and time-saving. However, the thermal post-treatment is generally required to obtain crystalline nanoparticles with a higher degree of crystallization.

There are only a few reported syntheses of UCNPs without thermal post-treatment. Co-precipitation solvents usually use surfactants, such as polyvinylpyrrolidone (PVP) and PEI, to control, stabilize, and surface functionalize nanoparticles. This process is easy to control and industrialize large-scale production. Yi et al. (2004), Wei et al. (2007) synthesized α- $NaYF_4$:Yb,Er nanoparticles in aqueous solution by using soluble rare earth compounds and fluorides as precursors and various organic solvents as ligands. Then calcined under the protection of a reducing atmosphere, the upconversion luminescence intensity of nanomaterials was significantly improved and β-$NaYF_4$:Yb,Er nanoparticles were obtained. Under the excitation of near-infrared light, the obtained α- $NaYF_4$:Yb,Er nanoparticles emit very weak upconversion luminescence, but after annealing at 400–600 °C, the luminescence intensity is increased by about 40 times.

Although the process and synthesis steps of the precipitation method are very simple, the cost required for the experiment is low, complicated and expensive synthetic equipments are not required, and its reaction conditions are mild, and no higher reaction temperature and no longer reaction time are required. The synthesized nanoparticles have excellent performance and are suitable for industrial production. However, there are still many drawbacks. The main disadvantage is that the directly obtained precipitated product has a low degree of crystallinity and nonuniform composition. Therefore, the obtained precipitated product requires further subsequent heat treatment such as calcination to enhance the crystallinity of the product. However, the subsequent heat treatment process can easily sinter the precipitated nanoparticles and often leads to severe agglomeration, which may damage the original grafted or coated organic ligands on the surface of the nanoparticles, which weakens the functionality and hydrophilicity of nanoparticles. It requires researchers further modify the surface of nanoparticles to achieve the purpose of functionalization, but it will further increase the particle size distribution and cannot control the size precisely and cannot be directly applied in biomedical field.

2.4.2 The Water/Solvothermal Method

The hydrothermal synthesis usually refers to a method in which the original mixture reacts in a closed system with an aqueous solution as the reaction medium under a certain temperature and water autogenous pressure. The difference between hydrothermal synthesis and solid-phase synthesis is that the "reactivity" is different. This difference is mainly reflected in the reaction mechanism. The mechanism of the solid phase reaction is mainly characterized by the interface, and the hydrothermal reaction is mainly characterized by the liquid phase reaction. Since the water is in a supercritical state under the condition of high temperature and high pressure, the physical and chemical properties of the substance in the water are greatly changed, so the hydrothermal chemical reaction is different from the normal state. Under hydrothermal conditions, the reactants are dissolved in various complexes and the water molecules themselves participate in this process. At present, people have a certain understanding of the nature of water and water solutions under hydrothermal conditions. The properties of water will mainly change as follows: (1) the vapor pressure becomes high; (2) the density becomes low; (3) the surface tension becomes low; 4) The viscosity becomes low; (5) The ion volume becomes large.

The hydrothermal preparation of materials has certain advantages. First of all, under hydrothermal conditions, water not only acts as a solvent, mineralizing agent, and medium for the transfer of pressure, but most of the reactants at high pressure can be completely or partially dissolved in water, thus facilitating the smooth reaction. Secondly, the hydrothermal process is affected by various reaction conditions, so it is possible to effectively control the reaction and crystal formation process by adjusting the precursor, mineralizer, reaction temperature, pressure and time, solution composition, PH value, and other factors. Third, the hydrothermal reaction conditions are mild, low energy consumption, and wide applicability, in addition to the preparation of nanomaterials, can also be used to prepare inorganic ceramic films and single crystals of larger size. Fourth, compared with other chemical methods, the raw materials required for hydrothermal reaction are cheap and easy to be obtained, and the reaction occurs in the fast reflux of the liquid phase. Therefore, the yield and purity of the obtained product is high, the phase of the product is relatively uniform. The crystal form of the product is easily realized. It is easy to control the particle size, dispersion, and morphology. Fifth, since the reaction is carried out in a sealed vessel, it is possible to provide suitable oxidation-reduction by controlling the reaction atmosphere to obtain a metastable phase that is difficult to obtain by other means. One of its characteristics is that the nonequilibrium thermodynamic synthesis chemistry is applied because the research system is generally in a nonideal and nonequilibrium state. Under high temperature and pressure conditions, water or other solvents are in a critical or supercritical state, and the reactivity is improved. The physical properties and chemical reaction properties of the substance in the solvent are greatly changed, so the thermochemical reaction of the solvent is different from the normal state. This method has become an important route for the synthesis of ultrafine particles, inorganic membranes, and single crystals. It has been reported that the use of hydrothermal method enables the synthesis of fluorides with almost zero oxygen content at very low temperatures, and it is expected that upconversion materials with relatively high conversion efficiency and relatively stable performance can be prepared.

For the hydrothermal synthesis of UCNPs, the typical preparation process is as follows: In the mixed system of water, ethanol, oleic acid, or sodium oleate, add rare earth ions and stir. Then fluoride aqueous solution was added and stirred evenly before hydrothermal treatment. The growth of rare earth luminescence materials can be controlled by adjusting the pH of the reaction system, the hydrothermal temperature and the hydrothermal treatment time. Professor Li Yadong's team (Liu et al. 2011; Chen et al. 2012b; Ohtsuki et al. 1997; Zhu et al. 1992; Boyer and van Veggel 2010) of Tsinghua University made outstanding contributions in this regard. They first added rare earth nitrate aqueous solutions to the microemulsion system of oleic acid or linoleic acid, water, ethanol, and sodium hydroxide. An aqueous solution of NaF was gradually added under stirring, and then transferred to an autoclave to adjust the reaction temperature and reaction time to control the morphology of the nanocrystals. They published academic papers in the journal Nature and introduced the liquid-solid-solution (LSS) phase transfer synthesis mechanism for hydrothermal systems. That is, ethanol and alkyl acids constitute the Liquid phase. The alkyl acid complexes of sodium alkyl and heavy metals form the Solid phase. The aqueous solutions of ethanol and heavy metals form the Solution phase. First, through ion exchange, heavy metal ions enter the Solid phase and substitute Na. The carboxylic acid complex of the alkyl chain is formed, and then it is reduced at the interface of the Liquid-Solid or Solution-Solid phase, and the alkyl chain is always wrapped around the periphery of nanoparticles and forms a hydrophobic structure at the surface of nanoparticles. When a nanoparticle grows to a certain size, it settles down due to gravity, so it is easy to collect nanoparticles at the bottom. They used this method to synthesize a series of morphology (spherical, cubic, rod-like, dendritic, etc.) and size (from a few nanometers to hundreds of nanometers) controllable and monodispersed rare earth luminescent materials with uniform particle size. Professor Zhao Dongyuan of Fudan University used the hydrothermal method to synthesize a series of beautiful UCNPs (mainly composed of $NaYF_4$) nanorods, nanotubes, and flower-like nanodisks. It was found that when the reaction temperature is less than 160 °C, the cubic phase $NaYF_4$ is mainly formed. With the increase of the reaction temperature and the prolongation of time, only $NaYF_4$ is dissolved and recrystallized into the hexagonal $NaYF_4$, i.e., transition only from the metastable $NaYF_4$ to the steady state β-$NaYF_4$, this process is irreversible.

Many other research groups have also used this method to prepare rare earth luminescence materials with various morphologies. Due to the large ionic radius of Gd, the formation phase is relatively stable. However, Y^{3+} has a relatively small ion polarization radius, β-$NaYF_4$ can only be produced under severe conditions, such as prolonged high-temperature hydrothermal conditions. The introduction of Gd^{3+} in $NaYF_4$ crystals can make $NaYF_4$ formation very fast and promote the rapid formation of β-$NaYF_4$. At the same time, because the upconversion luminescence efficiency has a strong dependence on the phase structure, it is also possible to adjust the upconversion luminescence efficiency of the UCNPs by changing the doping concentration of Gd^{3+}.

The solvothermal method is similar to the hydrothermal method. A high-boiling organic solvent such as oleic acid/octadecene and oleic acid/oleylamine is used as a solvent. Inorganic rare earth salts and fluorides are reaction raw materials. Then nanocrystals with a high degree of crystallization are synthesized at high temperatures by this method. This method can obtain uniform, controllable, monodispersed oil-soluble nanoproducts with a high crystallinity but without thermal post-treatment.

The reaction of water/solvothermal synthesis method is simple and inexpensive. The synthesized nanoparticles have a high degree of crystallinity, good dispersion, small, and controllable particle size with a narrow distribution, and do not need high-temperature treatment. The disadvantage is that the method has more influence factors on the particle size of the morphology, the reaction cannot be traced in real time in the sealed reaction vessel, the reaction temperature is high and there are potential safety hazards.

2.4.3 Thermal Decomposition Method

Thermal decomposition method is one of the methods that can effectively synthesize UCNPs with controlled morphology, monodisperse, and high fluorescence intensity. It is usually carried out in the environment of isolating oxygen and water by adding a metal-organic compound as a precursor in a high-boiling organic solvent. A rapid thermal decomposition reaction of rare earth element organic acid salts occurs at a relatively high synthesis temperature, and then upconversion luminescence nanoparticles are obtained. The thermal decomposition chemical reaction usually occurs in a mixed solvent consisting of two or more different solvents. Different solvents have different effects: the non-coordinating solvent in the mixed solvents provides a high-temperature condition and a sufficient energy for the rapid nucleation and crystal transformation of nanoparticles, and the coordination solvent can be adsorbed, grafted, or coated on the surface of the nanoparticles, so that the particles can be uniformly nucleated and prevented from agglomerating, then the dispersibility of the particles is enhanced. For example, a typical process for the thermal decomposition of rare earth salts of trifluoroacetic acid is to prepare various rare earth salts of trifluoroacetic acid first, then add them to high boiling organic solvents (oleic acid/octadecene, oleic acid/oleylamine, neat oleylamine, etc.). The temperature is raised to 340 °C under nitrogen protection to thermally decompose the rare earth trifluoride rare earth salts to produce rare earth fluoride nanomaterials. The high boiling organic solvents can also be heated to 250–340 °C and then the rare earth fluorides are injected into the trifluoroacetate. The solution is pyrolyzed to prepare a rare earth fluoride nanomaterial.

Professor Yan Chunhua's group of Peking University first used this method to prepare monodispersed LaF_3 triangular nanodisks (Zhang et al. 2005). Then, by changing the ratio of organic solvents, temperature, heating time, and other conditions, various rare earth luminescence materials with controllable morphology and different phase structures were prepared (Mai et al. 2006) and the synthesis mechanism of this method was studied. They believe that in different organic solvent systems, the nucleation free energy of rare earth ions is different, which directly determines the phase structure of UCNPs. For example, in the pyrolysis system of oleic acid/octadecene for Pr and Nd, $NaREF_4$ rods are easily formed; however, in oleic acid/oleylamine/octadecene pyrolysis system, the nucleation free energy is lower than the energy barrier of α-$NaREF_4$ and α-$NaREF_4$ can be formed. As long as oleylamine is present, the nucleation free energies of all rare earth ions are relatively low, and α-$NaYF_4$ and β-$NaYF_4$ can be formed at the same time. Both of these phases are relatively stable. However, β-$NaREF_4$ is more easily formed under high-temperature reaction conditions. For example, Professor Chow's group at the National University of Singapore (Yi and Chow 2006) prepared monodispersed UCNPs with uniform phase structure and a high upconversion luminescence efficiency 320 °C by using pure oleylamine system. Mai et al. (2006), Yi and Chow (2006) and Boyer et al. (2006) synthesize Re-UCNPs nanoparticles with a cubic host $NaYF_4$ by a thermal decomposition method. These high-quality nanoparticles are controllable in morphology and have good dispersion. The nanoparticles of β-$NaYF_4$ with a hexagonal phase are further synthesized through subsequent treatment or improved reaction conditions, and the synthesized nanoparticles show good upconversion luminescence properties.

Although the nanoparticles synthesized by the thermal decomposition method have a controllable particle size with a narrow size distribution, a good crystallinity and regular morphology, and monodispersed, there are toxic organic molecules adsorbed on the surface of the obtained nanoparticles, which makes the particles show obvious hydrophobicity and cannot bind the water molecules well. At the same time, the thermal decomposition method has disadvantages such as high cost, harsh reaction conditions, high requirement for raw materials, complicated reaction steps and processes, and a large number of toxic reaction byproducts.

2.4.4 The Sol-Gel Method

The sol-gel method is a kind of wet chemistry methods. It reacts in a solution containing metallic inorganic compounds or organic compounds. By adding special additives into the solution, the sol is formed through hydrolysis and polycondensation and then agglomerated (or obtaining by decondensation) Gel) to form gel. The gel is dried and heat treated to form the target crystal. This method is mainly used to prepare thin films and vitreous materials doped with upconversion luminescence nanoparticles, and usually requires subsequent heat treatment at high-temperature to increase the upconversion luminescence efficiency. Salas et al. (2005) synthesized ZrO_2:Yb,Er nanoparticles by modified sol-microemulsion-gel method.

The use of metal alkoxides or halides as precursors after hydrolysis, the condensation process results in nanoparticles, which are mainly used to prepare thin films and vitreous materials doped with UCNPs. In order to improve the upconversion luminescence efficiency, high-temperature calcinations are usually required. However, the particle size of the synthesized particles by this method is not controllable, and there is a serious agglomeration phenomenon after high-temperature calcination treatment, which is not suitable for further application in the biomedical field. Because of the central chemical point of this synthetic route is that the reactant molecules (or ions) are hydrolyzed (alcoholized) and polymerized in water (alcohol) solutions. For example, molecular-polymer-sol-gel-crystalline (or Amorphous), some specific structural and aggregated solid compounds or materials can be synthesized by effective control of their chemical processes.

2.4.5 Other Methods

Microwave-assisted heating is a green synthesis method that has been rapidly developed in recent years. The principle of microwave heating is an internal heating process, which uses the material molecules to absorb the electromagnetic energy in the microwave magnetic field, which generates heat energy by vibration at a high speed of several billion times so as to achieve the purpose of heating. The microwave-assisted heating method has the advantages of high heating speed, uniform heating, etc. In addition, the method can also reduce the activation energy of the reaction, thereby increasing the reaction rate. The microwave heating method for the synthesis of nanomaterials can greatly reduce the reaction time and reduce energy consumption. Schafer et al. (2010) synthesized orthorhombic RbY_2F_7:Yb,Er upconversion nanoparticles in the high-boiling solvent N-(2-hydroxyethyl) ethylenediamine by using the microwave-assisted heating method with NH_4F as the fluorine source. Based on XRD characterization, they calculated the average particle size to be about 60 nm (Wang et al. 2010c). $NAYF_4$:Er,Yb (A is Na or Li) upconversion nanoparticles were synthesized by Wang et al. by using microwave-assisted heating method and the morphology of the nanoparticles were adjusted by the concentration and composition of the reactants.

In addition, some new methods or traditional method improvements are constantly being explored and researched to obtain more stable and efficient rare earth upconversion nanoparticles.

2.5 Surface Properties and Cytotoxicity of Rare Earth Upconversion Luminescence Nanoparticles

The surface modification of UCNPs can not only improve the luminescence properties of nanocrystals, but also provide a potential application platform for various biological applications of UCNPs. In order to enable UCNPs to be used as biological fluorescence probe materials, methods for surface passivation or core-shell structures as shown in Fig. 2.2 should be developed to increase their upconversion luminescence efficiency and stability.

The concentration of doped ions on the surface of UCNPs nanoparticles is relatively high. These ions on the surface have a high surface energy, and easily interact with solvent ligands and surface impurities, which easily dissipates energy from absorbed photons or quench luminescence by non-radiative transition. Therefore, the upconversion luminescence efficiency of nanoparticles is lower than that of the corresponding bulk materials. Surface passivation is to coat the surface of UCNPs with a layer of passivation such as silicon dioxide layer, polymer layer or homogeneous rare earth ion doped host to form a core-shell structure as shown in Fig. 2.3. The surface particles are protected from being disturbed or oxidized, thereby effectively reducing the energy loss of the UCNPs and improving the upconversion luminescence efficiency. After the $NaYF_4$:Yb,Tm nanoparticle surface was modified with a layer of $NaYF_4$ and polyacrylic acid, which was about 2 nm thick and contained no other doped particles, its upconversion luminescence was nearly 30 times higher than the original material (Yi and Chow 2007). The $NaYF_4$:Yb, Er nanoparticle was modified by the same method as described above, and the upconversion luminescence efficiency was nearly 7 times higher than that of the original material. After the passivated the surface of KYF_4:Yb, Er and $NaGdF_4$:Yb, Er nanoparticles, the upconversion luminescence efficiency has also been significantly improved (Schäfer

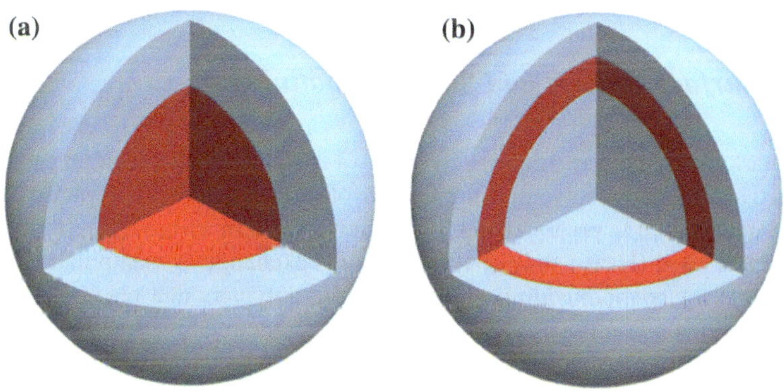

Fig. 2.2 Schematic diagram of a core-shell structure of a rare earth upconversion luminescence nanomaterials. The red layer contains luminescent lanthanide ions. **a** Standard core-shell structures with lanthanide ions are confined to the core layer. **b** Sandwich structure with lanthanide ions—Insertion into the middle layer (Chen et al. 2015)

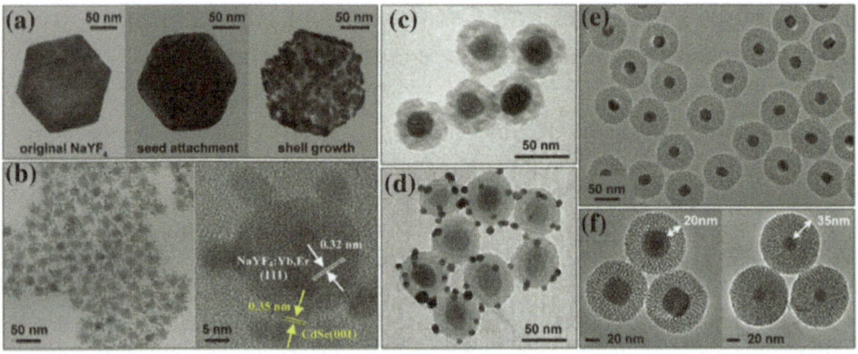

Fig. 2.3 TEM images of typical non-epitaxially upconverted core-shell nanoparticles. **a** $NaYF_4$:Yb,Tm@Au heterostructure constructed by Huang and Dua et al. **b** $NaYF_4$:Yb,Er@CdSe nanoparticles prepared by Rosei and Perepichka et al. **c** Nanoparticles encapsulated by Li and Zhang et al. in a silica shell on $NaYF_4$:Yb,Er. **d** Li and Xiong et al. coupled gold nanoparticles with $NaYF_4$:Yb,Er nanoparticles using a silica layer. **e** Porous silica coatings of Li and Lin et al. on $NaYF_4$:Yb,Er@ $NaGdF_4$:Yb nanoparticles. **f** $NaYF_4$:Yb,Tm nanoparticles prepared by Bu and Shi et al. (Li et al. 2008, 2011, 2013; Yan et al. 2010; Liu et al. 2012; Zhang et al. 2010)

et al. 2007). Hydrophobic Re-UCNPs nanoparticles were synthesized by thermal decomposition method, and then their surface were silanizated or coated with homogeneous rare earth doped host layers to form core-shell structure to obtain hydrophilic nanoparticles (Yang et al. 2011). However, the coating thickness of this shell layer has an effect on the upconversion luminescence efficiency. In addition, the hydrophilicity of the nanoparticles synthesized in the organic solvent or in the aqueous solution is poor. Therefore, in order to apply UCNPs biologically, the hydrophilic modification of the surface, such as addition of carboxyl, amino or aldehyde groups, is very important and necessary. This content will be discussed in detail in the next section.

2.5.1 Surface Groups and Surface Coatings

As biomarkers, UCNPs are required to bind to a specific biological target and can be dispersed in an aqueous solvent. In the absence of surface treatment, oil phase UCNPs cannot be dispersed in polar solvents. To achieve these functions, three approaches are generally used (Haase and Schafer 2011): (1) exchange or manipulation of surface organic ligands; (2) addition of amphiphilic polymers that react with surface nonpolar groups; (3) surface silanization. The most common method for improving dispersibility is to coat the surface of NPs with SiO_2 shells. In order to link these nanoparticles with biomolecules, reactive functional groups such as amine groups or carboxylic acid groups are needed. On the one hand, the dispersibility of nanoparticles in water can be improved, and on the other hand, it is ensured that covalent bonds can easily be passed with streptavidin.

Due to the large specific surface area of nanoparticles, the luminous efficiency of nanoparticles is generally lower than that of their bulk materials. In lanthanide upconversion nanomaterials, surface ligands with high-energy vibrational modes, such as OH or NH_2 groups, induce quenching of the excited state through multiphonon relaxation (Huang et al. 2014). If the dopant doping concentration is relatively high, the luminescence center inside the particle may further reduce the luminescence efficiency by the energy transfer from the neighboring ion to the surface. One of the main ways to reduce these energy losses is to coat the nanoparticles with a suitable shell material. In order to avoid energy loss, the energy transfer from the nanoparticle core to the shell material should be avoided. Therefore, the shell material generally uses the same undoped host material or a wide bandgap semiconductor material. With the same host, the lattice mismatch between the nanoparticle and the shell material is smaller. In some cases, the shell material is also doped with rare earth ions to improve upconversion luminescence properties or color controllability. For example, when the core is doped with Tm^{3+} and the shell is doped with Er^{3+}, various colors of tunable emission can be achieved, and the luminescence intensity is also apparent enhanced as compared with the core-shell structured nanoparticles without the ion-doped shell layer. The use of $NaGdF_4$: 20%Yb^{3+}, 2%Er^{3+}@$NaGdF_4$: 20%Yb^{3+} core-shell structure, the emission loss caused by the non-radiative transition of the core Er^{3+} was avoided, and on the other hand, the absorption of near-infrared light by the shell can transfer energy to the core as shown in Fig. 2.4. Then upconversion luminescence intensity was enhanced (Vetrone et al. 2009).

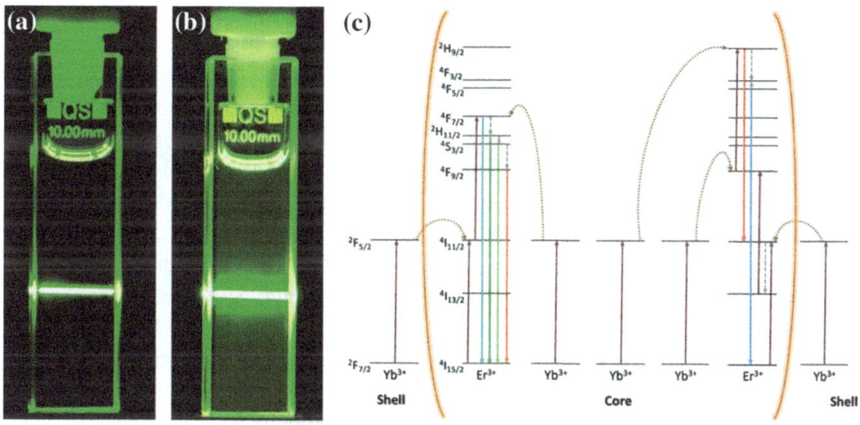

Fig. 2.4 **a**, **b** the $NaGdF_4$:Yb^{3+}, Er^{3+} active core @$NaGdF_4$ inert shell and $NaGdF_4$.Yb^{3+}, Er^{3+} active core@$NaGdF_4$:Yb^{3+} active shell nanoparticles colloidal solution formed in toluene was excited by 980 nm laser. **c** Illustration of the energy levels of Er^{3+} and Yb^{3+} doped ions and the mechanism of upconversion luminescence. Where the arrow represents energy transfer: the upward arrow represents energy absorption, the downward full arrow represents visible light emission, and the dotted arrow represents multiphonon relaxation (non-radiative decay)

2.5.2 Cytotoxicity

Upconversion nanoparticles (UCNPs) can emit visible light under low-density near-infrared excitation light, which avoids the interference of fluorescence signals generated by the biological material itself. And the infrared excitation is harmless to the organism and has a deep penetration depth. Therefore, it has unique advantages in the field of biomarkers. At the same time, the smaller size of UCNPs is also suitable for dispersing in biomolecules and macromolecules, and UCNPs has the advantages of narrow emission spectrum, adjustable luminescent color, good light stability, and low toxicity. However, the large specific surface area of nanoparticles also brings some problems. For example, the presence of surface defects makes the upconversion luminescence efficiency generally relatively low compared to bulk materials. On the other hand, the application of nanoparticles in biology also needs to consider the binding of nanoparticle surfaces with biomolecules and the influence of nanoparticles on biomolecules or organisms.

The application of UCNPs in the biological field will also be included in the use of organisms. Therefore, whether UCNPs contain harmful components to organisms, their cytotoxicity, has also become an important measure of UCNPs. Based on assessment of cell morphology and mitochondrial function (MTT and MTS assays) to detect cytotoxicity, it was found that lanthanide metal ion-doped nanoparticles are non-toxic to many cell lines. For example, Liu et al. (2010) tested the biocompatibility of SiO_2-coated Yb/Er-doped rare earth fluoride nanoparticles on the sebacic acid molecular layer and found that the nanoparticles were at a concentration of 800 mg/L. After incubation for 20 h, there was essentially no effect on the viability of human nasopharyngeal epidermal cancer KB cells. Prasad et al. (Liang et al. 2011) studied Tm^{3+} and Yb^{3+}-doped $NaYF_4$ nanocrystals on human pancreatic cancer cells and found that no significant cytotoxicity was found based on studies of MTS cell activity assays after phagocytosis of nanoparticles. Shan et al. (2008) also found that carboxy and amino-functionalized nanoparticles have limited or no effect on human osteosarcoma cells after 9 days of incubation.

The investigators also evaluated the in vivo cytotoxicity of upconversion nanocrystals. Abdul Jalil and Zhang (2008) intravenously injected SiO_2-coated hexagonal $NaYF_4$ upconversion nanoparticles into healthy mice. At a dose of 10 mg/kg body weight, no abnormal behavior was observed in mice. They found upconversion nanoparticles can be cleared most of the time after being injected into mice for 7 days. At the same time, through the MTT and LDH (lactate dehydrogenase) assays, they also studied the effect of different concentrations (from 1 to 100 mg/ml.) on cytotoxicity. Bone marrow stromal cells and skeletal muscle cells had strong tolerance to upconversion nanoparticles, nanoparticles into the cell did not cause serious damage to the cell membrane. In order to ensure the safety of the upconversion nanoparticles on organisms, it is necessary to study the influence of the size, shape, and surface properties of the upconversion nanoparticles on the cells over a longer period of time.

2.6 The Main Challenges of Rare Earth Upconversion Luminescence Materials

2.6.1 *Host Stability*

In general, the upconversion emission of a certain rare earth ion depends mainly on the properties of the host. The choice of materials requires not only low lattice vibrational energy of the host but also high chemical stability and mechanical strength of the host. In current materials, rare earth fluoride glass, especially ZrF_4 based glass, is an ideal material not only because of its low phonon energy (compared to oxide glass), but also because of its wide light transmission range and easier being to form waveguides and optical fibers. Considering the phonon energy, chlorides, bromides, and iodide glasses all have good application values. $CdCl_2$-based glasses, for example, have a phonon energy of about 250 cm^{-1}, which is very low compared to fluoride glasses, so that efficient upconversion luminescence can be obtained, and these materials are easily vitrified. However, these materials have poor chemical stability, mechanical strength, and thermal stability, which bring great difficulties to practical applications.

2.6.2 *Efficiency*

How to improve the luminous efficiency has always been the focus in the domain. After years of research, people have achieved certain results: The efficiency of upconverting red light has reached to 1%, green light 4%, and blue light 2%. From the foregoing, the upconversion luminescence efficiency is related to the choice of host materials, rare earth ions, and their concentrations. In a certain concentration range, the upconversion efficiency increases with the increase of rare earth ion concentration, and if it exceeds a certain concentration range, the concentration quenching occurs. However, in general, due to glass instability, it is difficult to synthesize upconversion glasses having a high concentration of rare earth ions (concentration of rare earth ions greater than 10 mol%). Although there are reports in the literature that fluorophosphate glass has good stability, it can be doped with a higher concentration of rare earth ions Pr^{3+}, and its spectrum can span the ultraviolet-near-infrared band. But an important issue is still the problem of its concentration annihilation.

2.6.3 Pump

Choosing the right path and the right pump source is critical to improving the efficiency of the upconversion laser. The rare earth ions are rich in energy levels, and each doping system corresponds to several pumping paths. In these pumping pathways, they should be used preferentially. At present, the commonly used upconversion laser pump source is infrared laser diodes. In the upconversion optical fiber waveguide laser, the pumping source and the rare earth ions have a larger working distance, and the pump energy loss is also large. In addition, the coupling between the pump source and the fiber is also a problem that cannot be ignored.

2.6.4 Narrow Range of Luminescence

At present, most of the upconversion systems are single-doped and double-doped systems. Although the emission of these systems can also be obtained with several wavelengths of emission, like Er^{3+} doping, 490, 552, and 662 nm emission can be obtained, their intensity is different, especially the blue part of the light is weak, almost submerged. And the obtained upconversion fiber laser has a tuning range of generally 10–30 nm. Therefore, it is imperative to obtain a material with a wide tuning range and wavelength range.

References

Abdul Jalil R, Zhang Y. Biocompatibility of silica coated $NaYF_4$ upconversion fluorescent nanocrystals. Biomaterials. 2008;29(30):4122–8.

Baldini F, Skovsen E, Homola J, et al. Photonics and microarray technology. 2007;6585:658516.

Bol AA, van Beek R, Meijerink A. On the incorporation of trivalent rare earth ions in II–VI semiconductor nanocrystals. Chem Mater. 2002;14(3):1121–6.

Boyer JC, van Veggel FC. Absolute quantum yield measurements of colloidal $NaYF_4$: Er^{3+}, Yb^{3+} upconverting nanoparticles. Nanoscale. 2010;2(8):1417–9.

Boyer JC, Vetrone F, Cuccia LA, et al. Synthesis of colloidal upconverting $NaYF_4$ nanocrystals doped with Er^{3+}, Yb^{3+} and Tm^{3+}, Yb^{3+} via thermal decomposition of lanthanide trifluoroacetate precursors. J Am Chem Soc. 2006;128(23):7444–5.

Cao C, Qin W, Zhang J, et al. Ultraviolet upconversion emissions of Gd^{3+}. Opt Lett. 2008;33(8):857.

Chen G, Liu H, Liang H, et al. Upconversion emission enhancement in Yb^{3+}/Er^{3+}-codoped Y_2O_3 nanocrystals by tridoping with Li+ Ions. J Phys Chem C. 2008;112(31):12030–6.

Chen G, Ohulchanskyy TY, Kumar R, et al. Ultrasmall monodisperse $NaYF_4$:Yb^{3+}/Tm^{3+} nanocrystals with enhanced near-infrared to near-infrared upconversion photoluminescence. ACS Nano. 2010;4(6):3163–8.

Chen G, Ohulchanskyy TY, Law WC, et al. Monodisperse $NaYbF_4$:Tm^{3+}/$NaGdF_4$ core/shell nanocrystals with near-infrared to near-infrared upconversion photoluminescence and magnetic resonance properties. Nanoscale. 2011;3(5):2003–8.

Chen G, Qiu H, Fan R, et al. Lanthanide-doped ultrasmall yttrium fluoride nanoparticles with enhanced multicolor upconversion photoluminescence. J Mater Chem. 2012a;22(38):20190.

Chen G, Ohulchanskyy TY, Liu S, et al. Core/shell $NaGdF_4$:Nd^{3+}/$NaGdF_4$ nanocrystals with efficient near-infrared to near-infrared downconversion photoluminescence for bioimaging applications. ACS Nano. 2012b;6(4):2969–77.

Chen X, Peng D, Ju Q, et al. Photon upconversion in core-shell nanoparticles. Chem Soc Rev. 2015;44(6):1318–30.

Dey R, Rai VK, Pandey A. Green upconversion emission in Nd^{3+}-Yb^{3+}-Zn^{2+}:Y_2O_3 phosphor. Spectrochim Acta Part A Mol Biomol Spectrosc. 2012;99:288–91.

Dong H, Sun LD, Yan CH. Basic understanding of the lanthanide related upconversion emissions. Nanoscale. 2013;5(13):5703–14.

Dong H, Sun LD, Yan CH. Energy transfer in lanthanide upconversion studies for extended optical applications. Chem Soc Rev. 2015;44(6):1608–34.

Ehlert O, Thomann R, Darbandi M, et al. A four-color colloidal multiplexing nanoparticle system. ACS Nano. 2008;2(1):120–4.

Guo H, Li Z, Qian H, et al. Seed-mediated synthesis of NaY F_4: Y b, Er/$NaGdF_4$ nanocrystals with improved upconversion fluorescence and MR relaxivity. Nanotechnology. 2010;21(12):125602.

Haase M, Schafer H. Upconverting nanoparticles. Angew Chem. 2011;50(26):5808–29.

Heer S, Kömpe K, Güdel HU, et al. Highly efficient multicolour upconversion emission in transparent colloids of lanthanide-doped $NaYF_4$ nanocrystals. Adv Mater. 2004;16(23–24):2102–5.

Huang P, Zheng W, Zhou S, et al. Lanthanide-doped $LiLuF_4$ upconversion nanoprobes for the detection of disease biomarkers. Angew Chem. 2014;53(5):1252–7.

Johnson NJJ, Oakden W, Stanisz GJ, et al. Size-tunable, ultrasmall $NaGdF_4$ nanoparticles: insights into their T1MRI contrast enhancement. Chem Mater. 2011;23(16):3714–22.

Joshi C, Rai SB. Structural, thermal, and optical properties of Pr^{3+}/Yb^{3+} co-doped oxyhalide tellurite glasses and its nano-crystalline parts. Solid State Sci. 2012;14(8):997–1003.

Krämer KW, Biner D, Frei G, et al. Hexagonal sodium yttrium fluoride based green and blue emitting upconversion phosphors. Chem Mater. 2004;16(7):1244–51.

Lezhnina MM, Jüstel T, Kätker H, et al. Efficient luminescence from rare-earth fluoride nanoparticles with optically functional shells. Adv Funct Mater. 2006;16(7):935–42.

Li Z, Zhang Y, Jiang S. Multicolor core/shell-structured upconversion fluorescent. Adv Mater. 2008;20(24):4765–9.

Li Z, Wang L, Wang Z, et al. Modification of $NaYF_4$:Yb,Er@SiO_2 nanoparticles with gold nanocrystals for tunable green-to-red upconversion emissions. J Phys Chem C. 2011;115(8):3291–6.

Li C, Yang D, Ma P, et al. Multifunctional upconversion mesoporous silica nanostructures for dual modal imaging and in vivo drug delivery. Small. 2013;9(24):4150–9.

Liang S, Liu Y, Tang Y, et al. A user-friendly method for synthesizing high-quality $NaYF_4$:Yb,Er(Tm) nanocrystals in liquid paraffin. J Nanomater. 2011;2011:1–7.

Liang Z, Wang X, Zhu W, et al. Upconversion nanocrystals mediated lateral-flow nanoplatform for in vitro detection. ACS Appl Mater Interfaces. 2017;9(4):3497–504.

Liu C, Chen D. Controlled synthesis of hexagon shaped lanthanide-doped LaF_3 nanoplates with multicolor upconversion fluorescence. J Mater Chem. 2007;17(37):3875.

Liu Q, Li C, Yang T, et al. "Drawing" upconversion nanophosphors into water through host-guest interaction. Chem Commun. 2010;46(30):5551–3.

Liu Q, Sun Y, Yang T, et al. Sub-10 nm hexagonal lanthanide doped $NaLuF_4$ upconversion nanocrystals for sensitive bioimaging in vivo. J Am Chem Soc. 2011;133(43):17122–5.

Liu J, Bu W, Zhang S, et al. Controlled synthesis of uniform and monodisperse upconversion core/mesoporous silica shell nanocomposites for bimodal imaging. Chemistry. 2012;18(8):2335–41.

Maciel GS, Guimar ESRB, Barreto PG, et al. The influence of Yb^{3+} doping on the upconversion luminescence of Pr^{3+} in aluminum oxide based powders prepared by combustion synthesis. Opt Mater. 2009;31(11):1735–40.

Mahalingam V, Vetrone F, Naccache R, et al. Colloidal Tm^{3+}/Yb^{3+} -doped $LiYF_4$ nanocrystals: multiple luminescence spanning the UV to NIR regions via low-energy excitation. Adv Mater. 2009;21(40):4025–8.

Mai HX, Zhang YW, Si R, et al. High-quality sodium rare-earth fluoride nanocrystals: controlled synthesis and optical properties. J Am Chem Soc. 2006;128(19):6426–36.

Mai H-X, Zhang Y-W, Sun L-D, et al. Highly efficient multicolor up-conversion emissions and their mechanisms of monodisperse $NaYF_4$:Yb,Er core and core/shell-structured nanocrystals. J Phys Chem C. 2007;111(37):13721–9.

Mart N-rodr Guez R, Valiente R, Polizzi S, et al. Upconversion luminescence in nanocrystals of $Gd_3Ga_5O_{12}$ and $Y_3Al_5O_{12}$ doped with $Tb^{3+}-Yb^{3+}$ and $Eu^{3+}-Yb^{3+}$. J Phys Chem C. 2009;113(28): 12195–200.

Ming C, Song F, Yan L. Spectroscopic study and green upconversion of Pr^{3+}/Yb^{3+} -codoped $NaY(WO_4)_2$ crystal. Opt Commun. 2013;286:217–20.

Naccache R, Vetrone F, Speghini A, et al. Cross-relaxation and upconversion processes in Pr^{3+} Singly Doped and Pr^{3+}/Yb^{3+} codoped nanocrystalline $Gd_3Ga_5O_{12}$: the sensitizer/activator relationship. J Phys Chem C. 2008;112(20):7750–6.

Nakazawa E. Cooperative optical transitions of $Yb^{3+}-Yb^{3+}$ and $Gd^{3+}-Yb^{3+}$ ion pairs in $YbPO_4$ hosts. J Lumin. 1976;12–13:675–80.

Ohtsuki T, Honkanen S, Najafi SI, et al. Cooperative upconversion effects on the performance of Er^{3+}-doped phosphate glass waveguide amplifiers. Journal of the Optical Society of America B. 1997;14(7):1838.

Ostrowski AD, Chan EM, Gargas DJ, et al. Controlled synthesis and single-particle imaging of bright, sub-10 nm lanthanide-doped upconverting nanocrystals. ACS Nano. 2012;6(3):2686–92.

Park YI, Kim JH, Lee KT, et al. Nonblinking and nonbleaching upconverting nanoparticles as an optical imaging nanoprobe and T1 magnetic resonance imaging contrast agent. Adv Mater. 2009;21(44):4467–71.

Patra A, Friend CS, Kapoor R, et al. Upconversion in Er^{3+}:ZrO_2 nanocrystals. J Phys Chem B. 2002;106(8):1909–12.

Patra A, Friend CS, Kapoor R, et al. Fluorescence upconversion properties of Er^{3+}-doped TiO_2 and $BaTiO_3$ nanocrystallites. Chem Mater. 2003;15(19):3650–5.

Pelle F, Dhaouadi M, Michely L, et al. Spectroscopic properties and upconversion in Pr^{3+} :YF_3 nanoparticles. Phys Chem Chem Phys. 2011;13(39):17453–60.

Qin X, Yokomori T, Ju Y. Flame synthesis and characterization of rare-earth (Er^{3+}, Ho^{3+}, and Tm^{3+}) doped upconversion nanophosphors. Appl Phys Lett. 2007;90(7):073104.

Ramakrishna PV, Pammi SVN, Samatha K. UV–visible upconversion studies of Nd^{3+} ions in lead tellurite glass. Solid State Commun. 2013;155:21–4.

Salas P, Angeles-Ch Vez C, Montoya JA, et al. Synthesis, characterization and luminescence properties of ZrO_2:Yb^{3+}–Er^{3+} nanophosphor. Opt Mater. 2005;27(7):1295–300.

Sarkar S, Meesaragandla B, Hazra C, et al. Sub-5 nm Ln^{3+}-doped $BaLuF_5$ nanocrystals: a platform to realize upconversion via interparticle energy transfer (IPET). Adv Mater. 2013;25(6):856–60.

Schäfer H, Ptacek P, Kömpe K, Haase M, et al. Lanthanide-doped $NaYF_4$ nanocrystals in aqueous solution displaying strong up-conversion emission. Chem Mater. 2007;19(6):1396–400.

Schäfer H, Ptacek P, Voss B, et al. Synthesis and characterization of upconversion fluorescent Yb^{3+}, Er^{3+} doped RbY_2F_7 nano- and microcrystals. Cryst Growth Des. 2010;10(5):2202–8.

Shan J, Chen J, Meng J, et al. Biofunctionalization, cytotoxicity, and cell uptake of lanthanide doped hydrophobically ligated $NaYF_4$ upconversion nanophosphors. J Appl Phys. 2008;104(9):094308.

Song W, Guo X, He G, et al. Ultraviolet upconversion emissions of Gd^{3+} in β-$NaLuF_4$:Yb^{3+}, Tm^{3+},Gd^{3+} nanocrystals. J Nanosci Nanotechnol. 2014;14(5):3722–5.

Su Q, Han S, Xie X, et al. The effect of surface coating on energy migration-mediated upconversion. J Am Chem Soc. 2012;134(51):20849–57.

Vetrone F, Boyer J-C, Capobianco JA, et al. Significance of Yb^{3+} concentration on the upconversion mechanisms in codoped Y_2O_3:Er^{3+}, Yb^{3+} nanocrystals. J Appl Phys. 2004;96(1): 661–7.

Vetrone F, Naccache R, Mahalingam V, et al. The active-core/active-shell approach: a strategy to enhance the upconversion luminescence in lanthanide-doped nanoparticles. Adv Funct Mater. 2009;19(18):2924–9.

Wang F, Liu X. Recent advances in the chemistry of lanthanide-doped upconversion nanocrystals. Chem Soc Rev. 2009;38(4):976–89.

Wang HQ, Nann T. Monodisperse upconverting nanocrystals by microwave-assisted synthesis. ACS Nano. 2009;3(11):3804–8.

Wang G, Peng Q, Li Y. Upconversion luminescence of monodisperse CaF_2:Yb^{3+}/Er^{3+} nanocrystals. J Am Chem Soc. 2009a;131(40):14200–1.

Wang Y, Tu L, Zhao J, et al. Upconversion luminescence of β-$NaYF_4$: Yb^{3+}, Er^{3+}@β-$NaYF_4$ core/shell nanoparticles: excitation power density and surface dependence. J Phys Chem C. 2009b;113(17):7164–9.

Wang F, Wang J, Liu X. Direct evidence of a surface quenching effect on size-dependent luminescence of upconversion nanoparticles. Angew Chem. 2010a;122(41):7618–22.

Wang F, Wang J, Liu X. Direct evidence of a surface quenching effect on size-dependent luminescence of upconversion nanoparticles. Angew Chem. 2010b;49(41):7456–60.

Wang H-Q, Tilley RD, Nann T. Size and shape evolution of upconverting nanoparticles using microwave assisted synthesis. CrystEngComm. 2010c;12(7):1993.

Wang F, Deng R, Wang J, et al. Tuning upconversion through energy migration in core-shell nanoparticles. Nat Mater. 2011a;10(12):968–73.

Wang J, Wang F, Wang C, et al. Single-band upconversion emission in lanthanide-doped $KMnF_3$ nanocrystals. Angew Chem. 2011b;50(44):10369–72.

Wang L, Lan M, Liu Z, et al. Enhanced deep-ultraviolet upconversion emission of Gd^{3+} sensitized by Yb^{3+} and Ho^{3+} in β-$NaLuF_4$ microcrystals under 980 nm excitation. J Mater Chem C. 2013;1(13):2485.

Wang J, Deng R, Macdonald MA, et al. Enhancing multiphoton upconversion through energy clustering at sublattice level. Nat Mater. 2014;13(2):157–62.

Wei Y, Lu F, Zhang X, et al. Synthesis and characterization of efficient near-infrared upconversion Yb and Tm codoped $NaYF_4$ nanocrystal reporter. J Alloy Compd. 2007;427(1–2):333–40.

Wong H-T, Vetrone F, Naccache R, et al. Water dispersible ultra-small multifunctional $KGdF_4$:Tm^{3+}, Yb^{3+} nanoparticles with near-infrared to near-infrared upconversion. J Mater Chem. 2011;21(41):16589.

Yan C, Dadvand A, Rosei F, et al. Near-IR photoresponse in new up-converting CdSe/$NaYF_4$:Yb,Er nanoheterostructures. J Am Chem Soc. 2010;132(26):8868–9.

Yang D, Li C, Li G, et al. Colloidal synthesis and remarkable enhancement of the upconversion luminescence of $BaGdF_5$:Yb^{3+}/Er^{3+} nanoparticles by active-shell modification. J Mater Chem. 2011;21(16):5923.

Yi GS, Chow GM. Synthesis of hexagonal-phase $NaYF_4$:Yb,Er and $NaYF_4$:Yb,Tm nanocrystals with efficient up-conversion. Adv Funct Mater. 2006;16(18):2324–9.

Yi G-S, Chow G-M. Water-soluble $NaYF_4$:Yb,Er(Tm)/$NaYF_4$/polymer core/shell/shell nanoparticles with significant enhancement of upconversion fluorescence. Chem Mater. 2007;19(3):341–3.

Yi G, Lu H, Zhao S, et al. Synthesis, characterization, and biological application of size-controlled nanocrystalline $NaYF_4$:Yb,Er infrared-to-visible up-conversion phosphors. Nano Lett. 2004;4(11):2191–6.

Yi G, Peng Y, Gao Z. Strong red-emitting near-infrared-to-visible upconversion fluorescent nanoparticles. Chem Mater. 2011;23(11):2729–34.

Zhang YW, Sun X, Si R, et al. Single-crystalline and monodisperse LaF_3 triangular nanoplates from a single-source precursor. J Am Chem Soc. 2005;127(10):3260–1.

Zhang H, Li Y, Ivanov IA, et al. Plasmonic modulation of the upconversion fluorescence in $NaYF_4$:Yb/Tm hexaplate nanocrystals using gold nanoparticles or nanoshells. Angew Chem. 2010;122(16):2927–30.

Zhang X, Yang P, Li C, et al. Facile and mass production synthesis of beta-$NaYF_4$:Yb^{3+}, Er^{3+}/Tm^{3+} 1D microstructures with multicolor up-conversion luminescence. Chem Commun. 2011;47(44):12143–5.

Zhang F, Che R, Li X, et al. Direct imaging the upconversion nanocrystal core/shell structure at the subnanometer level: shell thickness dependence in upconverting optical properties. Nano Lett. 2012;12(6):2852–8.

Zheng W, Zhou S, Chen Z, et al. Sub-10 nm lanthanide-doped CaF_2 nanoprobes for time-resolved luminescent biodetection. Angew Chem Int Ed. 2013a;52(26):6671–6.

Zheng W, Zhou S, Chen Z, et al. Sub-10 nm lanthanide-doped CaF_2 nanoprobes for time-resolved luminescent biodetection. Angew Chem. 2013b;125(26):6803–8.

Zhu C, Lu X, Zhang Z. Upconversion fluorescence of TeO_2PbO-based oxide glasses containing Er^{3+} ions. J Non-Cryst Solids. 1992;144:89–94.

Chapter 3
Modification and Functionalization of Up-Converting Phosphor Particles

Changqing Lin and Honggang Zhang

Abstract This chapter introduced several methods for modification and functionalization of the inert surface of up-converting phosphor particle (UCP), including siliconization of UCP surface and functionalization of the SiO_2 layer by carbodiimide (EDC), mixed anhydride (MA), N-hydroxysuccinimide (NHS), Glutaraldehyde and N-succinimidyl-3-(2-pyridyldithio) propionate (SPDP) methods. They work efficiently for preparing UCP-bioactive molecule conjugates.

Keywords Up-converting phosphor particle · UCP · Siliconization · Functionalization

3.1 Introduction

Up-converting phosphor Technology (UPT) is a new labeling technology developed based on up-conversion phosphor particle (UCP). UCP is a kind of up-converting nanoparticles (1–100 nm) composed of several rare earth metal elements (lanthanide or actinide)-doped transition metals, which usually emit photon with higher energy by photon up-conversion through absorbing two or more incident photons with relatively lower energy (https://en.wikipedia.org/wiki/Upconverting_nanoparticles). Due to its unique structure, UCP can be excited by infrared light to emit visible light. This process follows the anti-Stokes principle to achieve up-conversion of energy (Corstjens et al. 2003; Kuningas et al. 2006). This special property makes UCP as a marker in bioanalysis, with no background interference, no self-quenching, and is suitable for multiple and quantitative analysis, which are superior to traditional fluoresce markers. UPT will be widely used in rapid immunoassay, genomics research, high-throughput drug screening, microarray, surgical tissue imaging, food safety detec-

C. Lin (✉)
Beijing Key Laboratory of POCT for Bioemergency and Clinic (No. BZ0329), Beijing 100071, People's Republic of China
e-mail: changqing.lin@hotgen.com.cn

H. Zhang
Beijing Hotgen Biotechnology Co., Ltd., Beijing Z-Park Daxing Biomedical Industry Base, No. 9 Tianfujie, Beijing 102600, China

R. Yang (ed.), *Principles and Applications of Up-converting Phosphor Technology*,
https://doi.org/10.1007/978-981-32-9279-6_3

tion, and biochemical warfare defense (Banerjee and Jaiswal 2018; Lim et al. 2006; Quesada-Gonzalez and Merkoci 2015; van De Rijke et al. 2001; Sutherland et al. 2016; Vikesland and Wigginton 2010; Zuiderwijk et al. 2003).

Conventional grinding and synthesis are two common methods for UCP particle preparation. The grinding process is simple in production process, and no additional requirements are imposed on the production equipment, and UCP particles having a diameter of about 40 nm can be obtained. The UCP particles are chemically inert, therefore, they should be chemically modified for further functionalization on the surface before any biological molecule's conjugation.

3.2 Siliconization of UCP Surface

UCPs containing different host lattice, absorbers (activator ions), and emitters (sensitizer ions) can be prepared by synthetic methods. The main matrix of UCP is generally oxysulfide, fluoride, gallate, silicate, etc., but the surface of these substances has no active groups that can be used to bind to the bioactive molecules, if the surface of UCP is covered with a thin, uniform layer of Si, this problem can be solved. The SiO_3 crystal has good light transmittance, does not affect the optical properties of UCP, can be connected to various biologically active molecules through functionalization, and is a good interface material. The surface siliconization can be easily achieved by hydrolysis of $Si(OC_2H_5)_4$ in an aqueous solution containing NH_3 (Kumar and Zhang 2009; Posthuma-Trumpie et al. 2012; Corstjens et al. 2005). The reaction equation is as follows:

$$Y_2O_2S{:}YtEr + Si(OC_2H_5)_4 + 2H_2O \rightarrow Y_2O_2S{:}YtEr - SiO_2 + 4C_2H_5OH$$

3.3 Functionalization of the SiO_2 Layer on UCP Surface

The active free radicals can be modified onto the SiO_2 layer of UCP by using the following derivatives of silylation reagents including $-NH_2$, –COOH, –OH, etc. The most commonly used reagent (Fig. 3.1) is the first one, which can modify $-NH_2$ group onto the SiO_2 layer.

3.4 Conjugation of Bioactive Molecules to the UCP Surface

The modified UCP can be widely linked to biologically active molecules, such as antibodies and oligonucleotides, through a bifunctional cross-linking agent via a reactive free radical (such as $-NH_2$) on its surface. UCP can be flexibly applied as a reporter to a variety of analytical fields after being functionalized.

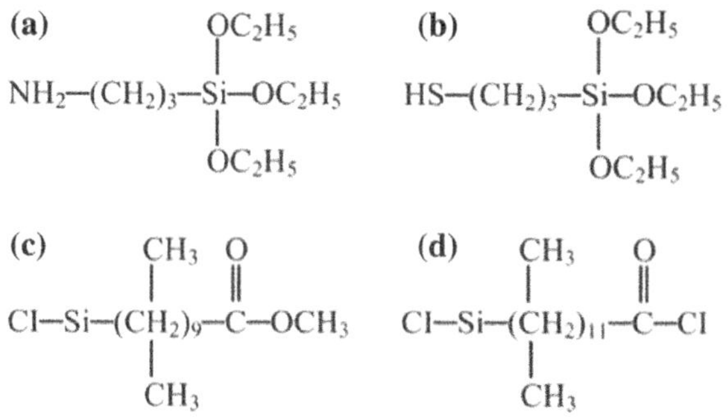

Fig. 3.1 The commonly used reagents for functionalization of the SiO_2 layer on UCP surface, **a** 3-Aminopropyltriethoxysilane; **b** 3-mercaptopropyltriethoxysilane; **c** monochlorodecanoyl-methoxy-dimethylsilane; **d** monochloro-lauroyl chloride- dimethylsilane

3.4.1 Cross-links Between Amino and Carboxyl

3.4.1.1 Carbodiimide (EDC) Method

The EDC method, originally proposed by Dr. Goodfriend, is the most commonly used strategy for conjugating small molecule haptens to protein carriers. EDC (R1N=C=NR2) is a chemically active reagent capable of forming an amide bond by dehydration condensation between an amino group and a carboxyl one. The free group (–COOH) modified on the surface of the UCP is first reacted with EDC to form an intermediate product, which is then reacted with an amino group of a biologically active material (such as an antigen, an antibody, etc.) to form a conjugate.

The chemical name of the frequently used aqueous EDC is 1-ethyl-3-(3-dimethylaminopropyl) carbodiimide (EDC), which can be combined with carboxyl groups on the UCP particles or bioactive materials. It can also be bound with its amino group. The optimal pH of the reaction is 5–9. The optimal pH should be selected according to the two substances cross-linked. For example, the environment of pH 7.0 is suitable for the coupling of most biological proteins.

The EDC method is very convenient. It only needs to dissolve the bioactive material and UCP particles in a proper ratio and mix well. Then, add EDC, stir the reaction at 4 °C or room temperature for 24 h, and finally separated by dialysis to remove the unbound portion for obtaining the coupled conjugate, UCP-bioactive material.

3.4.1.2 Mixed Anhydride (MA) Method

The free group –COOH modified on the surface of UCP reacts with isobutyl chloroformate in the presence of tri-n-butylamine or triethylamine to form a mixed acid anhydride; the intermediate product is mixed with an acid anhydride easily with the surface of the bioactive material. The group (–NH_2) reacts to form an amide bond.

3.4.1.3 N-Hydroxysuccinimide (NHS) Method

The free group –COOH modified on the surface of the UCP can be reacted with N-hydroxysuccinimide to form an activated ester, which is then coupled to a group ($-NH_2$) on the surface of the bioactive material.

3.4.2 Crosslinks between amino groups

3.4.2.1 Glutaraldehyde Method

Glutaraldehyde is a homobifunctional cross-linking agent. Its two aldehyde groups can form a Schiff base (–N=C–) with the amino group of two amino compounds, and a five-carbon linked bridge between the two compounds will be formed. This method can be used to link the bioactive materials to the UCP, and the addition of glutaraldehyde to the solution containing the UCP particles and the bioactive materials will promote the formation of a conjugate between UCP particles and bioactive materials.

3.4.2.2 SPDP Method

N-succinimidyl-3-(2-pyridyldithio) propionate (SPDP) for cross-linking or labeling between proteins has been widely used (Girshovich et al. 1995; Carlsson et al. 1978). SPDP is a heterobifunctional cross-linking agent that is different from the cross-linking agent of the glutaraldehyde.

This reaction mechanism is divided into 4 steps:

First, SPDP reacts with the free amino group ($-NH_2$) modified on the surface of the UCP, and introduces a protecting thiol group to obtain an intermediate product A;
Second, SPDP reacts with amino group of the biologically active material (such as IgG), then, the same reaction as the first step is performed for introducing a protective thiol group to obtain the intermediate product B;
Third, removing the dithiol pyridine protecting group in A and B with a reducing agent dithiothreitol (DTT) to obtain sulfur-containing A and B;
Last, between the sulfur-containing A and the thiol-containing pyridine protecting group, a combination of a thiol group and a disulfide bond forms a conjugate between UCP and a bioactive material.

3.4.2.3 Other Bifunctional Cross-Linkers

Bifunctional cross-linking agents such as toluene diisocyanate, benzoquinone, and fluorodinitrophenyl sulfone can also be used for cross-linking two amino-containing compounds of bioactive materials and UCP particles.

3.5 Development of UPT-based Immunochromatographic Platform

Beijing Hotgen Biotechnology Co., Ltd. has successfully developed UPT-based immunochromatographic platform based on the physical structure of classical gold immunochromatographic strip. The UCP particles are different from the particle size, density, and surface properties of colloidal gold particles. All solid phase materials related to immunochromatography (sample Pad, bond pad, analytical membrane, and absorbent pad) and chemical reaction system (surfactant, coating agents, pH, and ionic strength, etc.) have been optimized for developing such a technical platform; at the same time, through large specific surface area and highly active surface modification, the conjugating efficiency of UCP nanoparticles and bioactive molecules is highly promoted. Finally, UPT immunochromatographic test strip is developed for quantitative and qualitative detection of the targets. On this platform, we select many optimal monoclonal antibodies and different optimal reaction conditions for producing the UPT-based immunochromatographic strips for clinical and other on-site applications (Hao et al. 2017; Hong et al. 2010; Hu et al. 2018; Hua et al. 2015; Yan et al. 2006; Zhang et al. 2014; Zhao et al. 2016a; Zhao et al. 2016b), with short reaction time, high sensitivity, specificity and accuracy, good stability, and no environmental pollution and radiation hazards.

References

Banerjee R, Jaiswal A. Recent advances in nanoparticle-based lateral flow immunoassay as a point-of-care diagnostic tool for infectious agents and diseases. Analyst. 2018;143(9):1970–1996.

Carlsson J, Drevin H, Axen R. Protein thiolation and reversible protein-protein conjugation. N-Succinimidyl 3-(2-pyridyldithio) propionate, a new heterobifunctional reagent. Biochem J. 1978; 173(3).723–737.

Corstjens PL, Li S, Zuiderwijk M, Kardos K, Abrams WR, Niedbala RS, Tanke HJ. Infrared up-converting phosphors for bioassays. In: IEE proceedings nanobiotechnology, vol. 152, no. 2, p. 64 72; 2005.

Corstjens PL, Zuiderwijk M, Nilsson M, Feindt H, Sam Niedbala R, Tanke HJ. Lateral-flow and up-converting phosphor reporters to detect single-stranded nucleic acids in a sandwich-hybridization assay. Anal Biochem. 2003;312(2):191–200.

Girshovich AS, Bochkareva ES, Todd MJ, Lorimer GH. On the distribution of ligands within the asymmetric chaperonin complex, GroEL14. ADP7. GroES7. FEBS Lett. 1995; 366(1):17–20.

Hao M, Zhang P, Li B, Liu X, Zhao Y, Tan H, Sun C, Wang X, Wang X, Qiu H, et al. Development and evaluation of an up-converting phosphor technology-based lateral flow assay for the rapid, simultaneous detection of Vibrio cholerae serogroups O1 and O139. PLoS ONE. 2017;12(6):e0179937.
Hong W, Huang L, Wang H, Qu J, Guo Z, Xie C, Zhu Z, Zhang Y, Du Z, Yan Y, et al. Development of an up-converting phosphor technology-based 10-channel lateral flow assay for profiling antibodies against Yersinia pestis. J Microbiol Methods. 2010;83(2):133–140.
Hu Q, Wei Q, Zhang P, Li S, Xue L, Yang R, Wang C, Zhou L. An up-converting phosphor technology-based lateral flow assay for point-of-collection detection of morphine and methamphetamine in saliva. Analyst. 2018;143(19):4646–4654.
Hua F, Zhang P, Zhang F, Zhao Y, Li C, Sun C, Wang X, Yang R, Wang C, Yu A, et al. Development and evaluation of an up-converting phosphor technology-based lateral flow assay for rapid detection of Francisella tularensis. Scientific reports. 2015;5:17178.
Kumar M, Zhang P. Synthesis, characterization and biosensing application of photon upconverting nanoparticles. In: Proceedings—society of photo-optical instrumentation engineers; 2009, p. 7188.
Kuningas K, Ukonaho T, Pakkila H, Rantanen T, Rosenberg J, Lovgren T, Soukka T. Upconversion fluorescence resonance energy transfer in a homogeneous immunoassay for estradiol. Anal Chem. 2006;78(13):4690–4696.
Lim SF, Riehn R, Ryu WS, Khanarian N, Tung CK, Tank D, Austin RH. In vivo and scanning electron microscopy imaging of up-converting nanophosphors in Caenorhabditis elegans. Nano Lett. 2006;6(2):169–174.
Posthuma-Trumpie GA, Wichers JH, Koets M, Berendsen LB, van Amerongen A. Amorphous carbon nanoparticles: a versatile label for rapid diagnostic (immuno) assays. Anal Bioanal Chem. 2012;402(2):593–600.
Quesada-Gonzalez D, Merkoci A. Nanoparticle-based lateral flow biosensors. Biosens Bioelectron. 2015;73:47–63.
Sutherland JS, Mendy J, Gindeh A, Walzl G, Togun T, Owolabi O, Donkor S, Ota MO, Kon Fat ET, Ottenhoff TH, et al. Use of lateral flow assays to determine IP-10 and CCL4 levels in pleural effusions and whole blood for TB diagnosis. Tuberculosis (Edinb). 2016;96:31–36.
van De Rijke F, Zijlmans H, Li S, Vail T, Raap AK, Niedbala RS, Tanke HJ. Up-converting phosphor reporters for nucleic acid microarrays. Nat Biotechnol. 2001;19(3):273–276.
Vikesland PJ, Wigginton KR. Nanomaterial enabled biosensors for pathogen monitoring—a review. Environ Sci Technol. 2010;44(10):3656–3669.
Yan ZQ, Zhou L, Zhao YK, Wang J, Huang LH, Hu KX, Liu HH, Wang H, Guo ZB, Song YJ, et al. Rapid quantitative detection of *Yersinia pestis* by lateral-flow immunoassay and up-converting phosphor technology-based biosensor. Sens Actuators B. 2006;119:656–663.
Zhang P, Liu X, Wang C, Zhao Y, Hua F, Li C, Yang R, Zhou L. Evaluation of up-converting phosphor technology-based lateral flow strips for rapid detection of Bacillus anthracis Spore, Brucella spp., and Yersinia pestis. PloS one. 2014; 9(8):e105305.
Zhao Y, Liu X, Wang X, Sun C, Wang X, Zhang P, Qiu J, Yang R, Zhou L. Development and evaluation of an up-converting phosphor technology-based lateral flow assay for rapid and quantitative detection of aflatoxin B1 in crops. Talanta. 2016a;161:297–303.
Zhao Y, Wang H, Zhang P, Sun C, Wang X, Wang X, Yang R, Wang C, Zhou L. Rapid multiplex detection of 10 foodborne pathogens with an up-converting phosphor technology-based 10-channel lateral flow assay. Sci Rep. 2016b;6:21342.
Zuiderwijk M, Tanke HJ, Sam Niedbala R, Corstjens PL. An amplification-free hybridization-based DNA assay to detect Streptococcus pneumoniae utilizing the up-converting phosphor technology. Clin Biochem. 2003;36(5):401–403.

Chapter 4
Upconversion Nanoparticles-Based Point-of-Care Testing Technology

Yong Zhao

Abstract Upconversion nanoparticles-based point-of-care testing technology (UPT-POCT) is novel rapid analysis system that uses upconversion nanoparticles (UCNPs) as detecting reporters. Combined with the well-established lateral flow (LF) platform, UPT-POCT assays are suitable for sensitive, quantitative, and multiplexing diagnosis and detections. In this chapter the principles and theory of various UPT-POCT methods for the detection of different analytes (pathogens, toxins, drugs, biomarkers, and DNA) are presented, as well as the approaches for multiplexing detections.

Keywords UCNPs · UPT-POCT assay · Multiplexing detection

4.1 Introduction

The development of upconversion nanoparticles (UCNPs) has provided a powerful tool to modern detection technologies. Upconversion nanoparticles-based point-of-care testing (UPT-POCT) methods are novel rapid analysis systems using UCNPs as detecting reporters. UCNPs are rare earth doped ceramic particles that possess the unique property of infrared upconversion. They are chemically stable core-shell structures modified with surface functional groups that are suitable reporters in bioassays. Compared with conventional reporters (e.g., colloidal gold, latex beads, and fluorescent chemicals), UCNPs reporters can be quantitatively detected and allow permanent excitation as they are not easily bleached. Moreover, because of the absence of autofluorescence upon infrared excitation in natural specimens, UCNPs reporters have significant potential for use in high sensitivity assays. It has been reported that UCNPs reporters are 10–100-fold more sensitive than assays using conventional reporter systems such as colloidal gold or colored latex beads (Hampl et al. 2001).

Y. Zhao (✉)
Beijing Key Laboratory of POCT for Bioemergency and Clinic (No. BZ0329), Beijing 100071, China
e-mail: zhaoyong179@139.com

Beijing Institute of Microbiology and Epidemiology, No. 20, Dongdajie, Beijing 100071, Fengtai, People's Republic of China

R. Yang (ed.), *Principles and Applications of Up-converting Phosphor Technology*,
https://doi.org/10.1007/978-981-32-9279-6_4

Owing to the UCNPs reporter characteristics, UPT-POCT methods have attracted considerable attention and have great potential in rapid on-site detections.

Throughout the previous decades, there has been a continual push to develop POCT diagnostics that allow for rapid testing at or near the site of patient or in first response situations (Luppa et al. 2011; Sturenburg and Junker 2009). As alternatives to conventional diagnostic assays performed in specialized laboratories, POCT diagnostics can be successfully applied at the primary care level and are particularly suitable for use in remote settings with poor or no laboratory infrastructure (Mabey et al. 2004). Among the most established POCT platforms, immunochromatographic assays, also called lateral-flow (LF) assays, are one of the most popular methods. LF assays have been successfully applied in various areas owing to their low cost and ease of use in a variety of settings, and can be carried out by nontechnical personnel (Syedmoradi et al. 2017). However, the search for increased detection sensitivity and the possibility of quantitative detection remains an ongoing challenge for conventional LF assays. In this respect, several of the unique features of UCNPs reporters have the potential to improve LF assays. Combining the advantages of the UCNPs and LF platforms, the UPT-POCT detection method can realize rapid, sensitive, and quantitative on-site detections.

Over the past few years, various UPT-POCT detections have been developed and successfully applied in clinical diagnosis (Corstjens et al. 2008), food safety (Zhao et al. 2016a; Liu et al. 2016), drug abuse (Niedbala et al. 2001), and even public bio-safety investigations (Zhang et al. 2014). Depending on the nature of the analyte of interest, UPT-POCT detection has different formats, including competitive assays (for low molecular weight compounds, such as toxins and drugs), sandwich assays (for antigens with several epitopes, such as bacteria and viruses), and hybridization assays (for DNA targets). Besides assays for single targets, UPT-POCT can also facilitate multiple detections with many approaches, such as coating multiple test lines in one strip, using UCNPs of multiple colors, or employing a multiple channel strip or disk. This chapter presents the principles and theory of these UPT-POCT methods as well as the approaches for multiplexing detections.

4.2 UPT-POCT Principles

4.2.1 UPT-POCT Elements

A typical UPT-POCT strip consists of four components (sample pad, conjugate pad, membrane, and absorbent pad) integrated into a sheet of laminating card (Fig. 4.1). Essential to the assay is the transportation of a liquid sample (containing the analyte of interest) along the strip thereby passing several zones where specific biological recognition elements (antibodies, antigens, nucleic acids, or aptamers) have been coated. (Posthuma-Trumpie et al. 2009) The liquid moves along the paper because of the capillary force of the porosity membrane, but to maintain a continuous flow

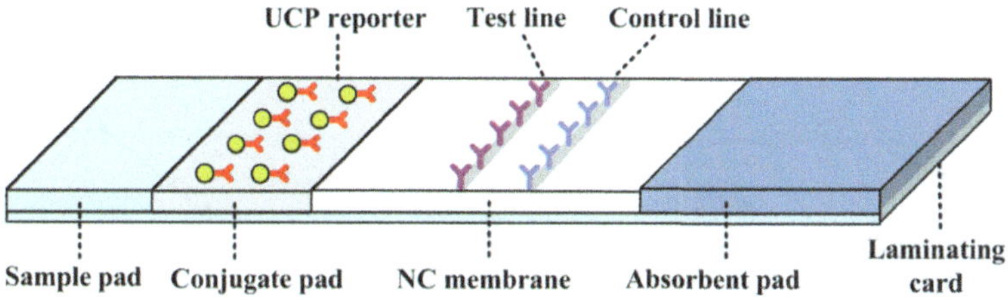

Fig. 4.1 Schematic description of the UPT-POCT strip. The strip is typically composed of a sample pad, a conjugate pad, a nitrocellulose membrane, an absorbent pad, and a laminating card. UCNPs reporters are immobilized in the conjugate pad. A pair of biological recognition elements (e.g. antibodies) are coated on the membrane as the test line and control line, respectively

an absorbent pad is required at the distal side of the strip. This absorbent pad can wick the liquid to the end of the strip, thus maintaining the flow. The other elements will be discussed in more detail below.

The sample pad where the liquid sample is applied is at one end of the strip and is usually made of cellulose or cross-linked silica. In close contact with the sample pad is the conjugate release pad, made of cross-linked silica. Reporters of UCNPs-labeled recognition elements (depending on the application) are sprayed and dried onto this pad. After addition of the sample, the UCNPs reporters will be dissolved and specific interactions will be initiated and will continue during the chromatographic process.

Currently, the most widely used strip membranes are produced from nitrocellulose (NC); others are nylon, polyethersulfone, polyethylene, or fused silica. At least two lines are sprayed onto the membrane: a test line and a control line. Recognition elements are coated at the test line and will specifically capture the target if the sample of analyte is positive, and a quantification response can be generated and collected. A response at the control line confirms a proper flow and can be used to reduce the variation of different strips. More test lines can be applied allowing for multiple analyte testing (Niedbala et al. 2001).

These paper materials are often thin and fragile. Therefore, the strip is attached to a plastic laminating card to allow cutting and handling. In addition, the strip is placed in a plastic holder, which exposes only a sample addition window and a reading window.

4.2.2 UPT-POCT Quantitative Analysis

Through the use of UCNPs reporters, UPT-POCT methods can acquire accurate quantitative analysis. The optical signals of UCNPs reporters on the strip can be collected by a UPT-POCT biosensor, which possesses a 980 nm laser and a transducer that can convert the optical signals to voltage signals (Yan et al. 2006; Mokkapati et al. 2007). As shown in Fig. 4.2a, the first peak (C) and the second peak (T) display the voltage units for the control line and the test line, respectively. The ratio Vt/Vc

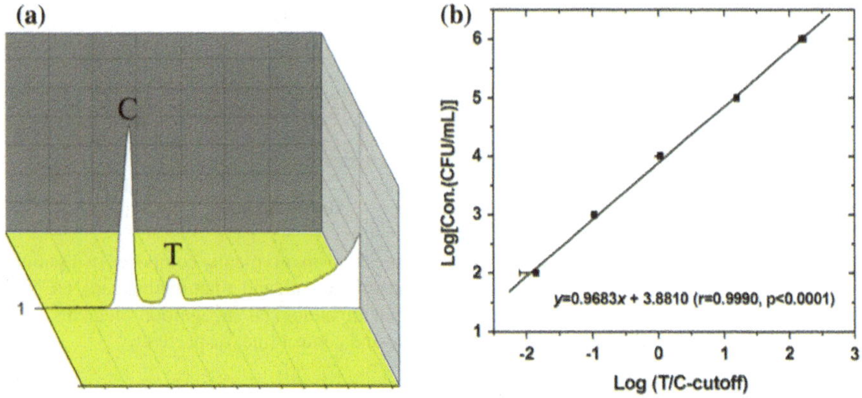

Fig. 4.2 Signal responses obtained by the UPT-POCT biosensor and the dose-response curve for quantification analysis. **a** The first peak (C) and the second peak (T) represent signals captured by the control line and the test line, respectively. The peak areas are auto-calculated by the biosensor, which represents the signal intensity, and the ratio Vt/Vc is regarded as the assay result. **b** Quantification analysis can be obtained according to the pre-defined dose-response curve, which is plotted with the ratio Vt/Vc of the samples on the x-axis and the corresponding concentrations on the y-axis

was obtained by the biosensor and was treated as the final result for each assay, where Vt represents the voltage for the test line signal, and Vc that for the control line signal.

The cut-off value of the assay is calculated as the mean ratio (Vt/Vc) of the blank controls plus (or minus) three standard deviations, which corresponds to 99% confidence. For the sandwich-format assay, samples with ratio Vt/Vc higher than the cut-off value (mean + 3 SD) are defined as positive and vice versa. For the competition-format assay, samples with ratio Vt/Vc lower than the cut-off value (mean − 3 SD) are defined as positive and vice versa.

To obtain a quantitative result, the dose-response curve is requisite. The curve can be obtained by testing serial dilutions of analyte samples, for example, bacterial samples with concentrations ranging from 10^2 to 10^6 CFU/ml. Then the dose-response curve can be plotted with the ratio Vt/Vc of the samples on the x-axis and the corresponding concentrations on the y-axis (Fig. 4.2b), to generate the corresponding equation of the curve. Unknown sample concentrations can then be quantified with the equation. To simplify the quantification analysis, information of the equation can be recorded in an RFID (Radio Frequency Identification) chip integrated into the strip (Zhang et al. 2014). Quantitative results for unknown samples can therefore be directly obtained by a UPT-POCT biosensor that can read the RFID information.

4.3 UPT-POCT Assay Formats

The UPT-POCT assay is designed to confirm the presence or absence of an analyte of interest, including bacteria, viruses, toxins, drugs, chemicals, and diagnostic biomarkers. There are several assay formats depending on the nature of the target analyte (Ngom et al. 2010). The two formats frequently used are the competitive assay and sandwich assay. In addition, DNA targets can be detected by the UPT-POCT assay in hybridization format when nucleic acid probes are used as recognition elements.

4.3.1 UPT-POCT Assay in Sandwich Format

For analytes with more than one epitope (such as pathogens, proteins, and some toxins), the sandwich format is applicable. This assay format employs two different antibodies that bind distinct epitopes of the analyte: a UCNPs-labeled antibody in the conjugate pad as a detector reagent and another antibody specific to the analyte in the test line as a capture reagent. An additional antibody specific to the detection antibody can be used in the control line.

When a liquid sample is applied to the sample pad, the liquid migrates by capillary force and the UCNPs-labeled antibody is released. Some of the detection antibody will bind to the analyte and some will remain free in the solution during the initial chromatographic process. Subsequently, the mixture passes through the test line, the analytes bind to the capture antibody and the excess free UCNPs-labeled antibody is captured by the control line. The response in the test line is directly proportional to the amount of analyte in the sample. Therefore, samples with higher Vt/Vc ratios are generally assumed to have higher concentrations. However, it should be noted that signals generated from the test line may be compromised if the concentration of the target analyte exceeds a certain critical value (Qian and Bau 2003), which could lead to inaccurate quantification.

4.3.2 UPT-POCT Assay in Competitive Format

The UPT-POCT assay in competitive format is employed most often when the test analyte is of low molecular weight or single epitope, such as aflatoxin B1 (AFB1) (Zhao et al. 2016b), N-sulfanilyl-4-aminobenzoic acid (Wang et al. 2007), methamphetamine, or other drugs (Niedbala et al. 2001). A typical scheme for the competitive design and the response is depicted in Fig. 4.3 (Zhao et al. 2016b). This assay format employs a UCNPs-labeled detection antibody against the analyte (e.g., monoclonal antibody from mouse against aflatoxin B1) and a UCNPs-labeled reference antibody (e.g., rabbit monoclonal or polyclonal antibody) in the conjugate pad. The test line in

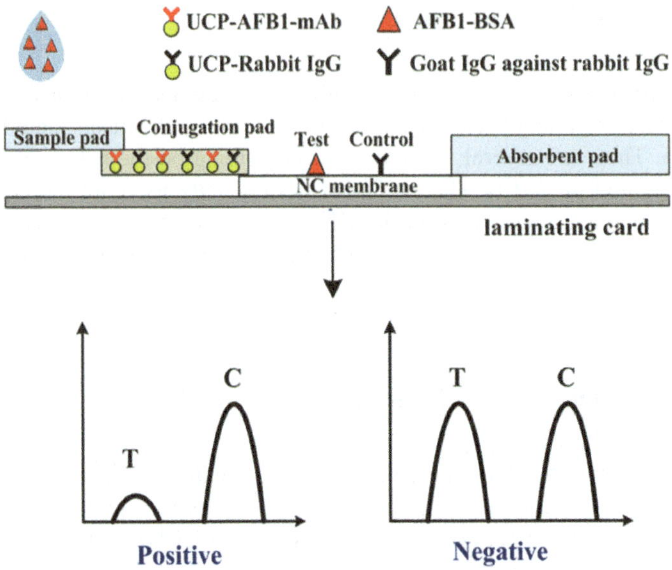

Fig. 4.3 Schematic description of the aflatoxin B1 (AFB1) UPT-POCT assay in competitive format (Zhao et al. 2016b). Typical results of the competitive immunoassay are presented, including positive and negative tests. The response (ratio *Vt/Vc*) is negatively correlated to the analyte concentration (i.e. more analyte present results in lower signal intensity; no analyte gives the highest signal intensity)

the membrane is coated with an analyte-protein conjugate (e.g., aflatoxin B1-BSA), and the control line is coated with antibodies against the reference antibody (e.g., goat IgG against rabbit antibody).

When liquid samples are positive for the analyte, the UCNPs-labeled detection antibody in the conjugate pad will first react with the target; then the free UCNPs reporters will bind to the test line (e.g., aflatoxin B1-BSA). With an increased amount of analyte in the sample, less free UCNPs reporters will be captured by the test line, which leads to a decrement in the "*Vt*" signal. For samples free of analyte, most of the free UCNPs reporters in the conjugate pad will flow along the paper and be captured by the test line, generating stronger "*Vt*" signals than those of the positive samples. Unlike the "T" signals, the "*Vc*" signals from the control line change little because of the invariable reaction between the UCNPs-labeled reference antibody and the antibodies against the reference antibody. Therefore, in competitive-format UPT-POCT assays the response (*Vt/Vc* ratio) is negatively correlated to the analyte concentration (i.e., more analyte present results in lower signal intensity; no analyte gives the highest signal intensity).

4.3.3 UPT-POCT Assay in Hybridization Format

DNA targets can be detected using the UPT-POCT assay in hybridization format (Zuiderwijk et al. 2003; Corstjens et al. 2003). This format can be set up in an antibody-dependent or antibody-independent manner. Both the set-ups can be designed for testing the presence or absence of a nucleic acid sequence specific to the target pathogen.

In the antibody-dependent format the analyte is double-stranded nucleic acid sequences (ds-amplicon) specific to the target pathogen, which are amplified using two PCR primers tagged with biotin and another tag (e.g., fluorescein isothiocyanate or digoxigenin) (Posthuma-Trumpie et al. 2009). Recognition of the analyte is achieved by binding to the UCNPs-avidin conjugate pre-sprayed in the conjugate pad, and subsequently captured by the tag-specific antibody (anti-fluorescein antibody) coated on the test line. In this format the response is directly proportional to the amount of analyte. An amplification step (PCR amplification) is often necessary, thus the use of good primers is important for the following assay.

For single-strand nucleic acid sequence (ss-amplicon) analytes, the antibody-independent format is applicable, which is based on the hybridization of nucleic acid sequences. This format employs two distinct complementary fragments of the target ss-amplicon: one labeled with the UCNPs reporter in the conjugate pad, and one sprayed in the test line as a capture probe. The immobilization of the capture probe on the nitrocellulose membrane can be performed through passive adsorption (Fig. 4.4c) or via passive absorption through a BSA-capture probe conjugate (Fig. 4.4b). The responses are directly proportional to the amount of analyte.

The following two points should be noted: (1) nucleic acid hybridization may take more time than antibody-antigen interaction; (2) it is very important to select good complementary nucleotide sequences that enable rapid and specific reactions.

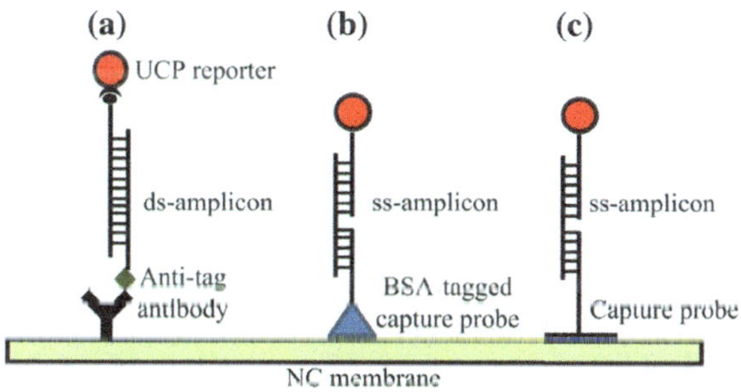

Fig. 4.4 Schematic description of the UPT-POCT assay in hybridization format (Ngom et al. 2010). **a** Antibody-dependent format with anti-tag antibody coated on the test line. **b** Antibody-independent format with the capture probe directly immobilized on the nitrocellulose membrane. **c** Antibody-independent format using BSA-capture probe conjugates immobilized on the nitrocellulose membrane

4.4 UPT-POCT Assay for Multiplexing Detection

4.4.1 UPT-POCT Multiplexing in One Strip

Multiplexing analysis of different targets is also feasible using the UPT-POCT method. Numerous multiplex UPT-POCT assays have been established by coating two or more test lines on one strip with common UCNPs or UCNPs with different optical spectra as reporters, as shown in Fig. 4.5a (Niedbala et al. 2001). The different test lines are placed separately (3 mm or more apart, depending on the number of test lines) and transversely across the strip such that the liquid sequentially passes the individual test lines. It has been reported that the nitrocellulose membrane of the test strip is long enough (4–5 mm wide and up to 8.5 cm long) to accommodate up to 12 lines 3 mm apart that could be distinguished by the test strip reader. In an example of a multiplexing UPT-POCT assay (Corstjens et al. 2007), antigens derived from HIV, hepatitis C virus (HCV), and tuberculosis were immobilized on three separated test lines on the membrane to capture respective antibodies. This assay is used for the multiplexed detection of human antibodies against the above pathogens, which could indicate the state of an infection.

To further enhance the degree of multiplexing, different types of UCNPS reporter can be used for different analytes. In an example (Niedbala et al. 2001), two types of UCNPS reporter were used for the detection of a panel of drugs (including amphetamines, methamphetamine, PCP, and opiates): one that exhibits blue emission at 475 nm, and a second that exhibits green emission at 550 nm. These two colors are completely spectrophotometrically distinguishable from each other with the band pass filters used in the reader. Antibodies for PCP and amphetamine were conjugated to green phosphors while methamphetamine and morphine were conjugated to blue-emitting phosphors. Analyzing the strip for each color phosphor, it was

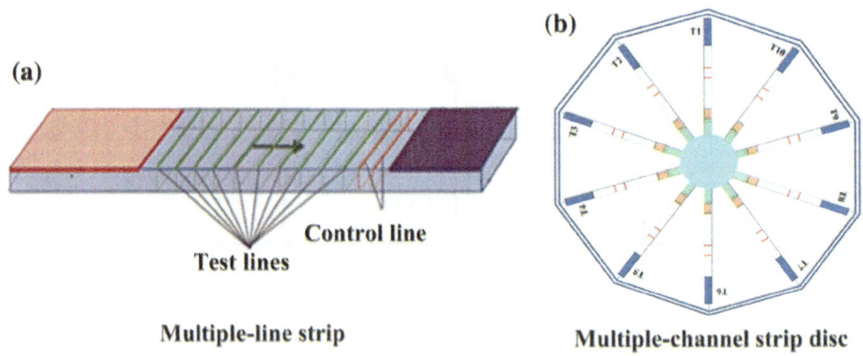

Fig. 4.5 UPT-POCT multiplexing assay in one strip or using a multi-channel disc. **a** Description of a UPT-POCT strip with up to 12 distinct capture lines on the membrane. **b** Schematic description of a 10-channel UPT-POCT disc holding 10 strips for different targets

possible to discriminate all drugs on the basis of phosphor color and position. The assay can be completed in 10 min, after which the result is scanned by a UPT reader.

It should be noted that the incorporation of multiple lines in a strip may raise several issues: (1) the nonspecific reaction or cross-reactivity may be complex, requiring careful development and optimization of the assay; (2) it may disturb the flow of UCNPS reporters to further downstream test lines, leading to high test signals in the test lines closest to the sample pad. Further studies are necessary to evaluate the maximum number of test lines that can be applied to a strip.

4.4.2 UPT-POCT Multiplexing in a Multiple-Channel Disc

Another approach to UPT-POCT multiplexing is to employ a multiple-channel disc, in which each channel holds a strip. Figure 4.5b shows a schematic description of a 10-channel UPT-POCT disc, which allows 10 UPT-LF strips to react simultaneously (Hong et al. 2010). The strips are overlapped at the sides of the sample pad. Besides conventional strip elements, a glass-fiber drainage piece is located in the disc center. By means of the symmetrical structure and the drainage piece, the samples applied through the sample-addition window can be distributed synchronously and uniformly into each strip, and thus this one-step assay can simultaneously detect multiple targets in one sample. In an example (Zhao et al. 2016a), a 10-channel UPT-POCT disc assay was developed for the simultaneous detection of ten foodborne pathogens in food samples. The disc holds ten double-antibody sandwich format UPT-LF strips with one in each channel for detecting the corresponding analytes. The results showed that solutions of multiple targets (10^7 CFU/mL for each bacteria) could be analyzed using the disc, and the 10 detection channels did not cross-react. The one-step multiplexing assay can be completed within 15 min of sample application, and quantitative analysis can be obtained within 2 min through the 10-channel UPT-POCT biosensor.

In contrast to the approach of multiple test lines in one strip, this technique can effectively reduce cross-reaction between the antibodies and targets. In addition, the disc configuration has the potential for use in high-throughput detections by setting more channels and using strips with two or more test lines in each channel. The challenge is to improve the sophistication of signal acquisition and miniaturize the instrument size toward POCT applications.

4.5 Future Outlook

On the basis of UPT technology and the LF platform, current UPT-POCT methods allow for efficient and cost-effective detection or diagnosis of an analyte or multiple analytes. However, there remain some challenges for the development of optimal POCT diagnostics, such as the improvement of analytical performance for ultra sensitive detection, and the development of home-care devices with sample handling and measurement capabilities. In addition, for commercial applications it is also important to improve the reproducibility of the tests, and good recognition elements (e.g., antibody, aptamer, and DNA/RNA) must be available and sufficient.

LF is currently the most suitable platform for POCT applications in resource-poor settings. With the development of technology, much of the recent work with UCNPs has focused on integrating this technology with microfluidics, which allows the miniaturization of POCT devices and automation of analysis processes. By integrating UPT-POCT into a lab-on-a-chip design, future UPT-POCT analytical systems with simple devices (such as smartphones) could replace complex laboratory assays, and have great potential to be powerful tools for personalized or home-care diagnostics.

References

Corstjens PL, Zuiderwijk M, Nilsson M, Feindt H, Sam Niedbala R, Tanke HJ. Lateral-flow and up-converting phosphor reporters to detect single-stranded nucleic acids in a sandwich-hybridization assay. Anal Biochem. 2003;312(2):191–200.

Corstjens PL, Chen Z, Zuiderwijk M, Bau HH, Abrams WR, Malamud D, Sam Niedbala R, Tanke HJ. Rapid assay format for multiplex detection of humoral immune responses to infectious disease pathogens (HIV, HCV, and TB). Ann N Y Acad Sci. 2007;1098:437–445.

Corstjens PL, van Lieshout L, Zuiderwijk M, Kornelis D, Tanke HJ, Deelder AM, van Dam GJ. Up-converting phosphor technology-based lateral flow assay for detection of Schistosoma circulating anodic antigen in serum. J Clin Microbiol. 2008;46(1):171–176.

Hampl J, Hall M, Mufti NA, Yao YM, MacQueen DB, Wright WH, Cooper DE. Upconverting phosphor reporters in immunochromatographic assays. Anal Biochem. 2001;288(2):176–187.

Hong W, Huang L, Wang H, Qu J, Guo Z, Xie C, Zhu Z, Zhang Y, Du Z, Yan Y, Zheng Y, Huang H, Yang R, Zhou L. Development of an up-converting phosphor technology-based 10-channel lateral flow assay for profiling antibodies against Yersinia pestis. J Microbiol Methods. 2010;83(2):133–140.

Liu X, Zhao Y, Sun C, Wang X, Zhang P, Qiu J, Yang R, Zhou L. Rapid detection of abrin in foods with an up-converting phosphor technology-based lateral flow assay. Sci Rep. 2016;6:34926.

Luppa PB, Müller C, Schlichtiger A, Schlebusch H. Point-of-care testing (POCT): Current techniques and future perspectives. TrAC Trends Anal Chem. 2011;30(6):887–898.

Mabey D, Peeling RW, Ustianowski A, Perkins MD. Diagnostics for the developing world. Nat Rev Microbiol. 2004;2(3):231–240.

Mokkapati VK, Sam Niedbala R, Kardos K, Perez RJ, Guo M, Tanke HJ, Corstjens PL. Evaluation of UPlink-RSV: prototype rapid antigen test for detection of respiratory syncytial virus infection. Ann N Y Acad Sci. 2007;1098:476–485.

Ngom B, Guo Y, Wang X, Bi D. Development and application of lateral flow test strip technology for detection of infectious agents and chemical contaminants: a review. Anal Bioanal Chem. 2010;397(3):1113–1135.

Niedbala RS, Feindt H, Kardos K, Vail T, Burton J, Bielska B, Li S, Milunic D, Bourdelle P, Vallejo R. Detection of analytes by immunoassay using up-converting phosphor technology. Anal Biochem. 2001;293(1):22–30.

Posthuma-Trumpie GA, Korf J, van Amerongen A. Lateral flow (immuno) assay: its strengths, weaknesses, opportunities and threats. A literature survey. Anal Bioanal Chem. 2009;393(2):569–582.

Qian S, Bau HH. A mathematical model of lateral flow bioreactions applied to sandwich assays. Anal Biochem. 2003;322(1):89–98.

Sturenburg E, Junker R. Point-of-care testing in microbiology: the advantages and disadvantages of immunochromatographic test strips. Dtsch Arztebl Int. 2009;106(4):48–54.

Syedmoradi L, Daneshpour M, Alvandipour M, Gomez FA, Hajghassem H, Omidfar K. Point of care testing: The impact of nanotechnology. Biosens Bioelectron. 2017;87:373–387.

Wang X, Li K, Shi D, Xiong N, Jin X, Yi J, Bi D. Development of an immunochromatographic lateral-flow test strip for rapid detection of sulfonamides in eggs and chicken muscles. J Agric Food Chem. 2007;55(6):2072–2078.

Yan Z, Zhou L, Zhao Y, Wang J, Huang L, Hu K, Liu H, Wang H, Guo Z, Song Y, Huang H, Yang R. Rapid quantitative detection of Yersinia pestis by lateral-flow immunoassay and up-converting phosphor technology-based biosensor. Sens Actuators B: Chem. 2006;119(2):656–663.

Zhang P, Liu X, Wang C, Zhao Y, Hua F, Li C, Yang R, Zhou L. Evaluation of up-converting phosphor technology-based lateral flow strips for rapid detection of Bacillus anthracis Spore, Brucella spp., and Yersinia pestis. PLoS One. 2014; 9(8):e105305.

Zhao Y, Wang H, Zhang P, Sun C, Wang X, Yang R, Wang C, Zhou L. Rapid multiplex detection of 10 foodborne pathogens with an up-converting phosphor technology-based 10-channel lateral flow assay. Sci Rep. 2016a;6:21342.

Zhao Y, Liu X, Wang X, Sun C, Zhang P, Qiu J, Yang R, Zhou L. Development and evaluation of an up-converting phosphor technology-based lateral flow assay for rapid and quantitative detection of aflatoxin B1 in crops. Talanta. 2016b;161:297–303.

Zuiderwijk M, Tanke HJ, Sam Niedbala R, Corstjens PL. An amplification-free hybridization-based DNA assay to detect Streptococcus pneumoniae utilizing the up-converting phosphor technology. Clin Biochem. 2003;36(5):401–403.

Chapter 5
Up-Converting Phosphor Technology-Based Biosensors

Huijie Huang, Lihua Huang and Yongkai Zhao

Abstract In this chapter, the basic concepts of the up-converting phosphor technology-based biosensor (UPT biosensor) was introduced, the working principle and the key technologies of the UPT immunoassay analyzer were discussed, and the analyzer development was presented. The UPT biosensor is an optical biosensor using up-converting phosphor (UCP) particles as the label, and consists of a UPT lateral flow (LF) strip and a UPT immunoassay analyzer. The analyzer mainly consists of the photoelectric measurement system, control system, algorithm module and mechanical structure. By measuring the upconversion luminescence signal distribution of UCP particles on the UPT LF strip, the analyzer obtains the target analyte concentration of the sample on the basis of the response characteristic curve. The prototype development and the productionization of analyzers were completed. The calibration results indicate that the series of UPT biosensors have good response linearity and the detection limit for the plague F1 antigen reaches the level of ng/ml and even sub ng/ml. The UPT biosensors have been successfully applied in the on-site detection of plague in plague foci, rapid on-site detection of anti-bioterrorism at the frontier port, major activity security and clinical diagnosis.

Keywords Optical biosensor · UPT biosensor · Response characteristics · UPT immunoassay analyzer · One-dimensionally confocal scanning (1-D confocal scanning)

The up-converting phosphor technology-based biosensor (UPT biosensor) is an optical biosensor that utilizes up-converting phosphor (UCP) particles as the label. A UPT biosensor usually consists of a UPT lateral flow (LF) strip and a UPT immunoassay analyzer. The analyzer is an intelligent photoelectric detection instrument that integrates optical, mechanical, electronic and algorithmic technologies.

Since detailed instructions for the principle and preparation of the UPT LF strip was provided in the previous chapter, the basic concepts of the UPT biosensor,

H. Huang (✉) · L. Huang · Y. Zhao
Shanghai Institute of Optics and Fine Mechanics, Chinese Academy of Sciences, Shanghai 201800, China
e-mail: huanghuijie@siom.ac.cn

R. Yang (ed.), *Principles and Applications of Up-converting Phosphor Technology*,
https://doi.org/10.1007/978-981-32-9279-6_5

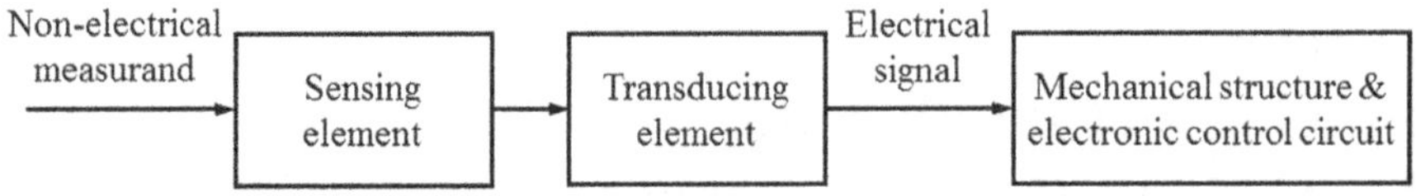

Fig. 5.1 Schematic of three elements of a sensor

working principle, key techniques of the UPT immunoassay analyzer, and analyzer development are discussed in this chapter.

5.1 Basic Concepts of UPT Biosensor

5.1.1 Sensor

According to the national standard of the People's Republic of China (*General Terminology for Transducers* GB/T 7665-2005), a transducer/sensor refers to a component or device that can sense a measurand and translate it into a usable output signal based on certain rules. A sensor is usually composed of a sensing element and a transducing element. The usable output signal is generally understood to be an electrical signal. The sensing element refers to the part of a sensor that can directly sense or respond to a measurand, which is also called the receptor. The transducing element, known as a transducer, refers to the part of a sensor that can translate the measurand responded by the sensing element into an electrical signal suitable for transmission or measurement. In addition, through corresponding mechanical structures and electronic control circuits, the function of the sensor can be extended to become more complete and more powerful. Sensing element, transducing element and mechanical structure and electronic control circuit are the three essential elements of a sensor (Fig. 5.1).

In the modern sense, the output signals of sensors are usually electrical signals or other forms of information that can be used for transmission, processing, storage, display, recording, control, etc. Therefore, sensors are the first link for realizing automatic detection and control, and they are precise detection instruments that involve material science, microelectronics, instrumentation technology, environmental science, physics, chemistry, biomedicine and bionics, computer technology and many other disciplines and technologies.

5.1.2 Biosensor

According to the national standard of the People's Republic of China (*General Terminology for Transducers* GB/T 7665-2005), a biosensor refers to a sensor that trans-

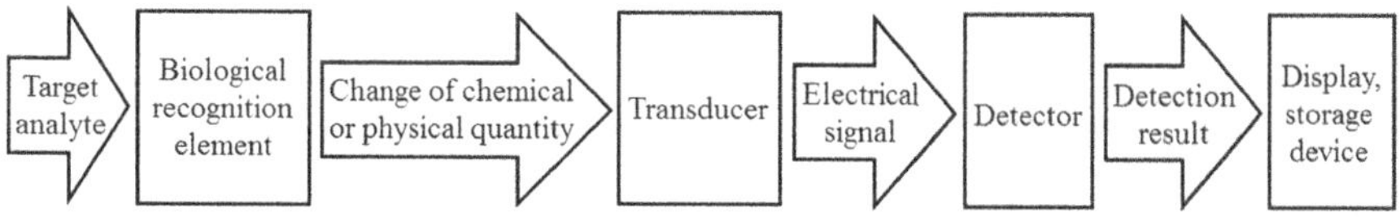

Fig. 5.2 Block diagram of working principle of biosensor

lates the characteristic quantities of a sensed measurand into usable output signals by utilizing the molecular recognition function of bioactive substances.

Specifically, biosensors are sensors that use bioactive units such as enzymes, antibodies, antigens, nucleic acids, hormones, or organisms themselves such as cells and organelles as the sensing elements and have high selectivity to target analytes. Sensing elements are also called biological recognition elements or molecular recognition elements, and transducing elements, i.e., transducers or signal converters, are called primary apparatuses. The working principle of a biosensor is shown in Fig. 5.2: the target analyte diffuses into the sensing element and causes biological reaction through molecular recognition. The information produced is then translated into an electrical signal which is quantifiable and processable by the corresponding physical or chemical transducer and processed and outputted by the detector (secondary apparatus). Accordingly, the properties and the concentration of the target analyte are obtained, and the purpose of the detection and analysis is achieved (Zhang 2006).

Biosensors mainly have two classification methods: the biological recognition element/molecular recognition element classification and the signal converter classification method (Zhang 2006). According to the biological/molecular recognition elements, biosensors can be divided into seven types: enzyme sensor, immunosensor or immunol sensor, microbial sensor, tissue sensor, organelle sensor, nucleic acid sensor (DNA/RNA biosensor) and molecular imprinted biosensor. Among these, the molecular recognition elements in the molecular imprinted biosensors belong to biological derivatives. According to the signal converters, biosensors mainly include the electrochemical biosensor or bioelectrode, optical biosensor, calorimetric biosensor or thermal biosensor, semiconductor biosensor, conductive/impedance biosensor, acoustic wave biosensor and cantilever biosensor.

The biosensor is state-of-the-art technology that has been developed with the mutual penetration of biological, chemical, physical, material, medical and electronic technologies, along with other disciplines. It has important application value in the fields of biomedicine, environmental protection, food safety and military medicine (Si 2003)

5.1.3 *Immunosensor*

An immunosensor is a biosensor that senses the amount of antigen or antibody and translates it into a usable signal. This biosensor imitates the natural immunoreactions of organisms and uses the recognition between an antigen and an antibody to detect molecules, cells and microorganisms (Miao 2005). The quality of an immunosensor depends on the selectivity and affinity of the combination between an antibody or an antigen and a target analyte (Si 2003).

Immunosensors have important theoretical research and practical application value. Generally, according to the categories of signal converters and the types of the translation signals, immunosensors can be divided into the following categories: (1) electrochemistry class: potential, current and conductance/impedance methods; (2) optics class: light absorption, photoluminescence (fluorescence or upconversion luminescence), chemiluminescence, interferometry and surface plasmon resonance methods; and (3) acoustic class: quartz crystal microbalance, surface acoustic wave and surface longitudinal wave methods.

Since the molecular binding between an antigen and an antibody does not involve a catalytic effect and amplification of an enzyme, direct detection has a low sensitivity and a long response time. Therefore, the immunosensor usually uses a variety of labeling methods to reveal and amplify the signals. As a result, the immunosensor is divided into two types: label and label-free. The main labeling methods are:

(1) microbead or nanoparticle labeling method: the labels include luminescent particles, gold nanoparticles, lattices, liposomes, magnetic particles and others. Some labels generate light signals (upconversion luminescence and fluorescence), some generate color variations, some increase in molecular weight, and some are enriched in the positions to be measured.
(2) molecule labeling method: this method uses small molecular labels and enzyme labels. Molecular labels directly generate optical signals (fluorescence and luminescence resulting from chemical reaction) or electrical signals (electroactive molecular markers), and enzyme labels produce optical or electrical signals under the actions of enzymes.

The labeling method may label either an antigen or an antibody (Byfield and Abuknesha 1994). For the antibody labeling method, when an immunoreaction occurs, the antigen under test, abbreviated as Ag, and the conjugate of the label and the sensitive antibody, represented by Ab*, forms the conjugate of label-sensitive antibody-antigen under test represented by Ab*-Ag (Fig. 5.3a). When the reaction ends, a corresponding relationship is established between the quantity of the Ag and the quantity of the label. Hence the quantity of an Ag can be obtained by detecting the quantity of the label. Similarly, the detection principle of antigen labeling method can be obtained (Fig. 5.3b).

5.1.4 Immunochromatographic Biosensor

The solid-phase membrane-based immunoassay with the membrane as the solid phase carrier of the immunoreaction is a new rapid in vitro diagnosis technique which has been developed and improved based on mature technologies including

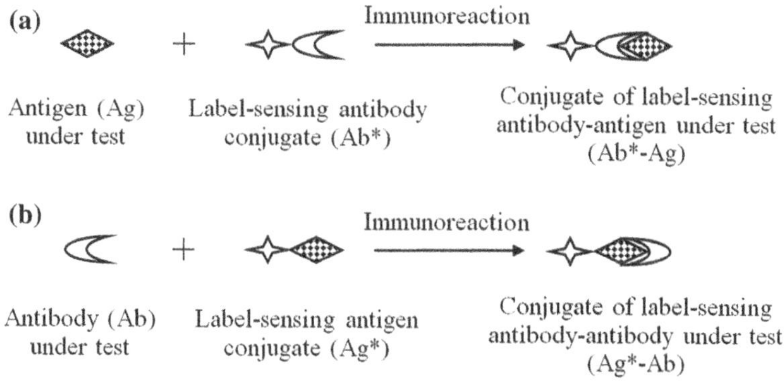

Fig. 5.3 Schematic of immunoreaction principle of antibody labeling method (**a**) and antigen labeling method (**b**)

enzyme-linked immunosorbent assay (ELISA), monoclonal antibody preparation, in vitro protein expression, label (colloidal gold, fluorescence particle and UCP particle) preparation and so on since the 1990s. According to the liquid-sample flow direction in the reaction process, this technique can be divided into two types: (1) lateral flow, also known as immunochromatographic assay, and (2) flow through, also known as immuno-filtration assay. Of the two, lateral flow is simpler and more rapid, and has high specificity and sensitivity. Therefore, it has been developed rapidly and plays an important role in clinical inspection, environmental protection, illicit drug monitoring, military medicine and other fields (Li and Yang 2003).

The immunochromatography assay generally uses a nitrocellulose membrane as the carrier. The liquid sample added on one end of the membrane strip is the mobile phase. Through the capillary action of the microporous membrane, the liquid sample containing the labels and the target analytes slowly laterally migrates to the other end, like lateral chromatography. In this migration process, the conjugates of the labels and the target analytes that have formed will bind to the receptors (sensitive antigens or antibodies) immobilized in a certain region on the membrane, and unrelated substances will cross the region and be separated. The detection results can then be determined through the indicating signals of the labels. The carrier of the immunochromatographic assay is usually called the lateral flow (LF) strip and acts as the sensing element and possesses part of the function of the transducing element through the bound labels.

The quantity of the labels conjugated to the immobilized target analytes on the LF strip can directly characterize the concentration of the target analytes in the liquid sample, and the concentration information can be obtained by the detection of the labels by the detection instrument. Accordingly, the LF strip and the detection instrument constitute a complete immunochromatographic biosensor. The performance of a biosensor is largely determined by the properties of the label because the information of the target analyte is indirectly obtained by measuring the label.

5.1.5 Optical Biosensor

An optical biosensor is a biosensor that involves the detection of optical signals. It uses optically sensitive components as the signal converter, reveals the selective and microscopic molecular recognition process by the optical signal that can be quantitatively detected, and consequently detects and analyzes the biological information by analyzing the change of the optical signal caused by the biological reaction.

The relationship between light and biological reactions can be summarized into three aspects: (1) bioluminescence reactions caused by the oxidation of special substances (such as fluorescein) in some organisms (creatures that glow include fireflies, some animals in the deep sea and some bacteria); (2) the transmission of light in biological substances, such as light absorption, light quenching and stimulated light emission of biological substances; and (3) the interference of biological substances to light propagation.

Biosensors using optical signals as detection indexes appeared in the 1980s. This kind of sensor has two main advantages: (1) the photoelectric detection device has high sensitivity and can improve the sensitivity of the corresponding biosensor; (2) the noise of the sensor is low because the transmission of the optical signal is not interfered with by the external electromagnetic field. Therefore, optical biosensors have received the most concentrated attention in the biosensor research field (Zhang 2006).

The optical phenomena involved in existing optical biosensors mainly include light absorption, light reflection, fluorescence, upconversion luminescence, chemiluminescence, bioluminescence, Raman scattering, photoacoustic and surface plasmon resonance (Table 5.1). The optical parameters involved in biological reactions mainly include intensity, phase, polarization and wavelength. Almost all the mechanisms of the interactions between light and a substance can be applied to the design of optical biosensors. The optical biosensor is a kind of modern optical instrument

Table 5.1 Main working principles of optical biosensor

Changes of optical properties	Light absorption	Light scattering
Intensity (amplitude)	Absorption/reflection spectrum	Mie scattering, Rayleigh scattering
Wavelength	Fluorescence, upconversion luminescence, luminescence spectrum	Raman spectroscopy
Time domain characteristics	Time resolved fluorescence, upconversion luminescence, luminescence spectrum	Dynamic light scattering spectrum
Phase/polarization state	Polarization absorption/circular dichroism spectrum/fluorescence anisotropy	Ellipsometry

that integrates biology, optics, precise mechanism, electronics, computer and other technologies (Huang et al. 2010).

Colorimetry and luminescence are the two common principles of optical biosensors. Colorimetry is mainly based on the changes in the color or the absorbance of products produced by an enzyme reaction or an immunoreaction. Luminescence mainly includes chemiluminescence or bioluminescence, fluorescence and upconversion luminescence. From luminous-substance sources, optical signals can be divided into self luminescence, label luminescence, enzyme reaction product luminescence and pyrolysis product luminescence.

Optical biosensors generate optical signals by the direct method or the indirect method. Direct-method optical biosensors usually do not need additional specific labels. Their signals depend only on the biological reactions, and their detection sensitivities are limited to nonspecific pollutants. Indirect-method optical biosensors use added labels to reveal biological reactions. These labels are usually fluorescent dyes or nanoparticles that can be optically detected. Therefore, in the indirect method, only the nonspecific pollutants which carry the used labels may generate serious background noise. Other nonspecific biological substances, as long as their optical properties do not overlap with the labels, cannot form large background noise. Accordingly, the detection-signal quality of the indirect method is superior to that of the direct method for complex and contaminated samples. Hence, the application of the indirect method is much more extensive than that of the direct method in practical optical biosensors.

According to whether carriers of biological reactions are optical elements, optical biosensors can be divided into two categories: (1) optical biosensors that use optical elements as carriers, which mainly include the optical fiber biosensor, planar optical waveguide biosensor and surface plasmon resonance biosensor; and (2) optical biosensors that use nonoptical elements as carriers, which mainly include the immunochromatographic optical biosensor, biochip scanner, flow cytometry, microfluidic chip system and chemiluminescence immunoassay system.

Optical biosensors are sensitive, specific, stable, provide high throughput, and are suitable for rapid on-site detection, and are widely used in biomedical diagnostics, environmental protection, anti-bioterrorism, food safety, drug screening, inspection and quarantine, drug monitoring and other fields.

5.1.6 UPT Biosensor

Immunochromatographic optical biosensors not only retain the characteristics of immunochromatography such as sensitive, fast, simple and multi-detection capability, but also have the advantages of optical detection technology, which include sensitivity, rapidity and non-destructiveness. However, as discussed in Sect. 5.1.4, the performance of immunochromatographic biosensors largely depends on the characteristics of the labels used.

The UPT biosensor is an immunochromatographic optical biosensor that uses UCP particles as the label and consists mainly of a UPT LF strip and a UPT immunoassay analyzer. The UCP is composed of a host matrix material and doped rare-earth ions. The doped rare-earth ions are divided into two species: absorber and emitter. The combinations of different absorbers, emitters and host matrices provide the UCPs with different optical properties. The UCP particles can conjugate with a variety of biologically active molecules after biological modification and activation and are thus used as biolabels (Zhou et al. 2003; Hampl et al. 2001; Niedbala et al. 2000, 2001).

In addition to the above advantages of the immunochromatographic optical biosensor, the UPT biosensor has the following prominent technical characteristics due to the use of UCP particle labels:

(1) high sensitivity: the upconversion luminescence (hereafter called the UCL) of UCP possesses high efficiency, and each emission peak of the luminescence has a large anti-Stokes shift and narrow width; therefore, the UCL is easy to be separated from the excitation light. In addition, the LF strip and other substances do not generate stray light under the infrared excitation. Hence, the detection sensitivity is high, and the background noise is low.
(2) high stability: the UCPs are inorganic inert materials with high chemical stability. The UCL of UCP particles is a purely physical process that generates within the materials. The materials have no decay or degradation in the luminescence process, and the luminescence stability is high. Therefore, generally, the UCL properties of the UCP particles are not affected by the environmental factors, and the detection processes are not affected by the liquid samples or sampling conditions.
(3) high flexibility: UCP particles with different compositions have diverse characteristic spectra which can lead to flexible quantitative detection and multi-detection by combining with biological technologies.
(4) high security: UCP particles have high stability and low potential toxicity, the infrared excitation light has low energy, and the detected upconversion luminescence is in the visible region. Therefore, the detection operations are less harmful to operators, the inspected person and the environment.

The UPT immunochromatography was discussed in detail in Chap. 4. In this chapter, the UPT immunoassay analyzer for the interpretation of a UPT LF strip is introduced.

5.2 Working Principle of UPT Immunoassay Analyzer

The UPT immunoassay analyzer is a photoelectric detection instrument for the quantitative interpretation of a UPT LF strip. The analyzer uses the LF strip after biological reaction as the detection object. By measuring the UCL signal distribution of the UCP particles on the LF strip, the analyzer obtains the target analyte concentration of the sample based on the response characteristic curve.

5.2.1 Quantitative Detection Principle of UPT LF Strip

The UPT LF strip is the carrier on which the immunoreaction of the target analyte in the liquid sample reacts and can be regarded as the sensing element of the UPT biosensor. After the reaction, the quantity of the UCP particles bound in the test line of the LF strip has a proportion relationship with the concentration of the target analyte.

A UPT LF strip (hereafter called an LF strip) is mainly composed of five parts: sample pad, conjugate pad, analytical membrane, absorbent pad and laminating card (Fig. 5.4) (Yan et al. 2006). It can be loaded into a disposable plastic cartridge (Fig. 5.5). At the time of detection, the sample is added into the sample pad of the LF strip and enter the conjugate pad by an infiltration and siphonic effect, which makes the fixed conjugates of the UCP particles and the bioactive molecules inside the pad

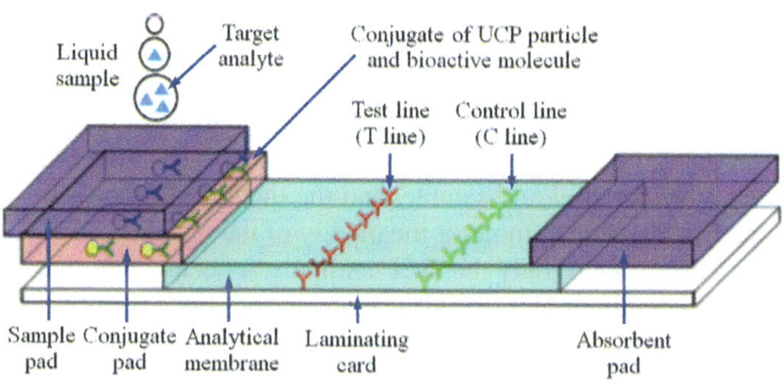

Fig. 5.4 Structure schematic of UPT LF strip. Modified from Lu et al. (2006)

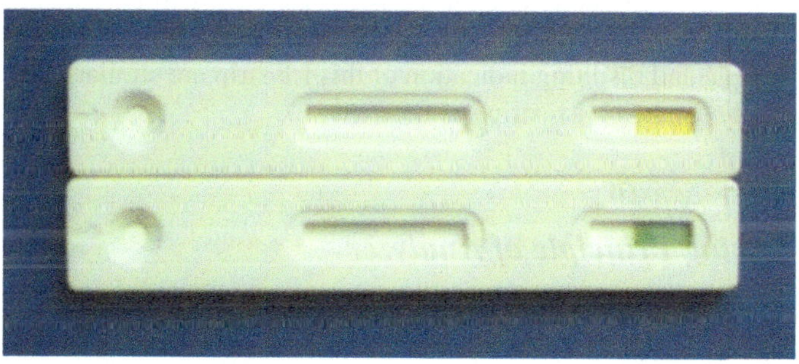

Fig. 5.5 Photograph of LF strips with disposable plastic cartridges (top of picture is LF strip before reaction, and bottom is LF strip after reaction)

redissolve and free, then leaves the conjugate pad by capillary action and siphonage of the absorbent pad, enters the analytical membrane, and flows in the direction of the absorbent pad. During this process, the conjugates of the UCP particles and the bioactive molecules, the target analytes and the receptors immobilized in the test line and the control line will undergo specific immunoreactions.

The immunoreactions on the LF strip can be categorized as sandwich mode, competitive mode and indirect mode. The double-antibody sandwich mode can be used to detect pathogens, microorganisms and macromolecular antigens in samples, and the double-antigen sandwich mode can be used to detect antibodies.

The antigen detection principle of a double-antibody sandwich LF strip is shown in Fig. 5.6. The conjugate of the UCP particle and the target analyte-specific antibody A (i.e., the antibody can specifically bind with the target analyte, but it can only bind to the A site of the target analyte), abbreviated as UCP-antibody A conjugate, is fixed in the conjugate pad of the LF strip. The target analyte-specific antibody B (i.e., the antibody can specifically bind with the target analyte, but it can only bind to the B site of the target analyte) is immobilized on the analytical membrane as the test line, and the second antibody that can bind with antibody A is immobilized on the membrane as the control line. An ending-index window on the absorbent pad provides a color change to determine that the liquid sample reaches the absorbent pad by passing though the analytical membrane.

For the positive sample, the UCP-antibody A-target analyte-antibody B conjugates are immobilized in the test line when the reaction ends, and the UCP-antibody A-second antibody conjugates are immobilized in the control line (Fig. 5.6a). The quantity of the UCP particles in the test line is related to the concentration of the target analyte in the added liquid sample, and the quantity of the UCP particles in the control line can be used to monitor the quality of the reaction.

For the negative samples, only the UCP-antibody A-second antibody conjugates are immobilized in the control line when the reaction ends, while only nonspecifically stranded UCP particles are in the test line (Fig. 5.6b). The quantity of the UCP particles in the control line can still be used to monitor the quality of the reaction.

If antibody A and antibody B on the LF strip are replaced with antigen A and antigen B, respectively, and the specific antibody A of antigen A is immobilized in the control line, the double-antigen sandwich LF strip will be obtained and can be used to detect the specific antibody caused by a certain pathogen infection. The reaction principle and resulting indication of this LF strip are similar to those of the double-antibody sandwich LF strip.

5.2.2 *Working Principle of Analyzer*

The UCL of UCP particles generally has multiple emission peaks within the ultraviolet, visible and near-infrared regions (Hampl et al. 2001). Within a certain range of the excitation light power (density), the relationship between the intensity I_{vis} of each UCL emission peak and the excitation light power (density) P is

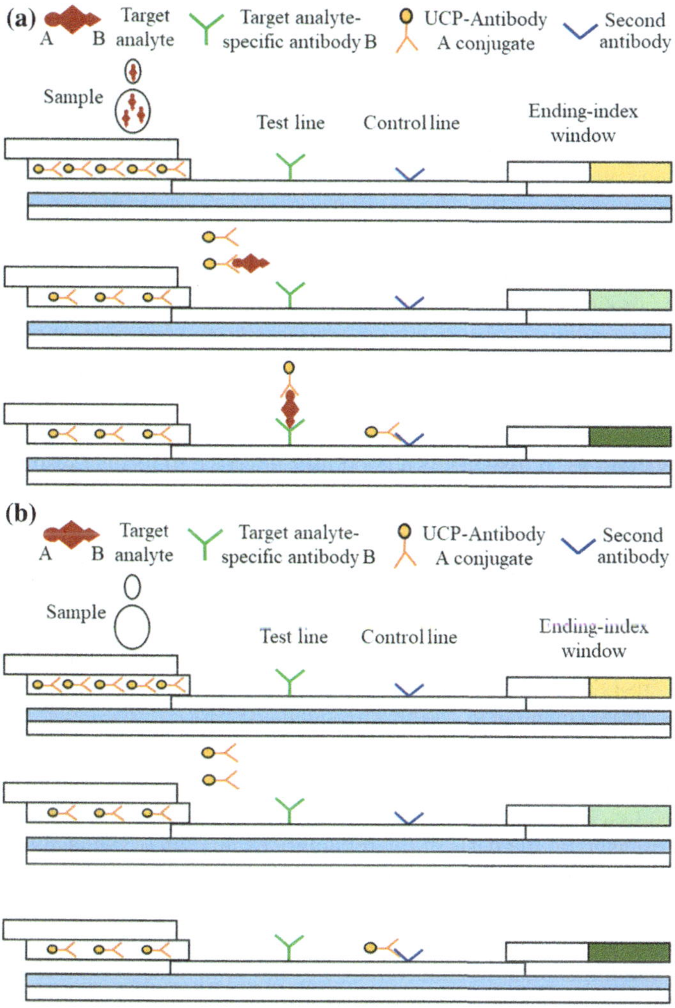

Fig. 5.6 Schematic of immunoreactions of positive sample (**a**) and negative sample (**b**) on double-antibody sandwich LF strips

$$I_{vis}(P) \propto P^n, \tag{5.1}$$

where n is the constant corresponding to each emission peak (Menyuk et al. 1972; Page et al. 1998; Suyver et al. 2005a, b, c, 2006; Zhao et al. 1999, 2000, 2005a).

An LF strip combining with UCP particles after a reaction can be approximately regarded as a plane Lambertian source. As the prepared UCP particles are almost uniform, their luminescence properties are similar under identical excitation light conditions. In the UPT immunoassay analyzer, when the UCL of the UCP particles within the uniformly illuminated region is collected by the optical system with an equal solid angle, the above relationship of Eq. (5.1) can be converted to the corresponding relationship between the optical signal collected by the optical system in the UCL of these UCP particles and the excitation light power. Therefore, there is a quantitative correspondence between the luminous flux of the UCL collected by the optical system and the quantity of UCP particles within the illumination region. When the subsequent photoelectric signal conversion system, signal processing system and other systems have linear characteristics, the amplitude of the ultimate output electric signal also has a linear correspondence with the quantity of measured particles. The above excitation light source, optical system, photoelectric signal conversion system and pre-signal processing system are combined and called the photoelectric measurement system.

To complete the detection automatically, intelligently and quickly, the analyzer needs a computer control system to control and coordinate the overall system and various subsystems to perform condition monitoring, signal acquisition, result output and other operations. It also needs a good user interface to receive instructions and display information.

In practical analyzer, using the quantity of UCP particles in the test line of the LF strip to directly characterize the target analyte concentration of the sample can introduce the measurement error that should be eliminated or reduced by corresponding technical means. Among them, using the ratio of the UCL signal in the test line to that in the control line of the same LF strip as the detection result can significantly reduce systematic errors and further improve the sensitivity and repeatability of detection (Zhou et al. 2003; Huang et al. 2009; Zhao et al. 2005b).

Finally, it is necessary to calibrate the analyzer. It is to determine the relationship between the target analyte concentration of the sample and the detection signal of the analyzer, i.e., the response characteristic curve or the standard working curve of the UPT biosensor, on the basis of the reliability evaluation of the analyzer.

Based on the above working principle, the UPT immunoassay analyzer first measures the UCL signal distribution of the UCP particles on the LF strip, then calculates the UCL signal in the test line (T line) (S_T) and that in the control line (C line) (S_C) (Fig. 5.7). The target analyte concentration of the sample according to the ratio of S_T/S_C and the response characteristic curve is then obtained.

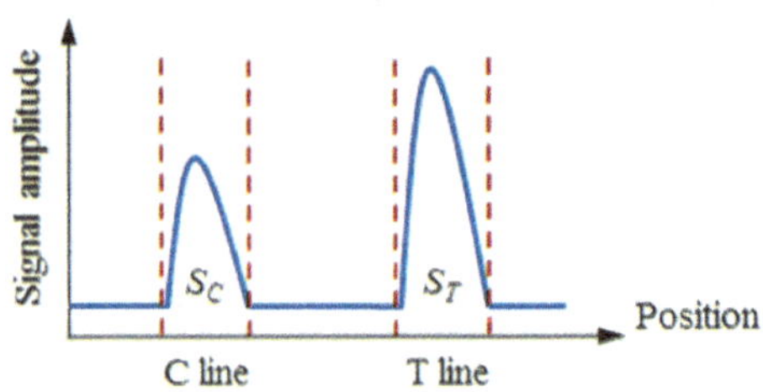

Fig. 5.7 Schematic of UCL signal distribution of UCP particles on LF strip

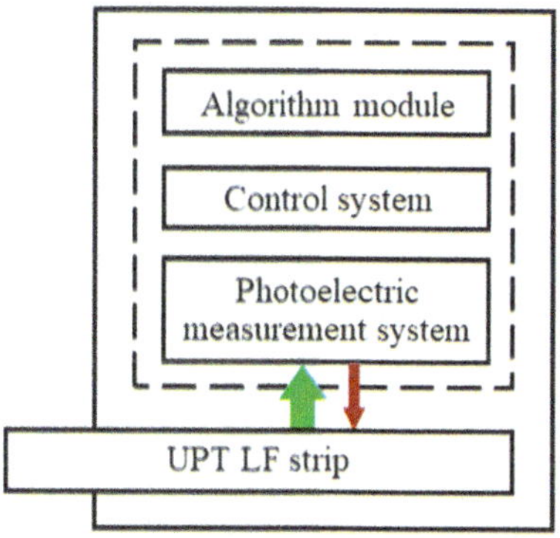

Fig. 5.8 Composition of UPT immunoassay analyzer

5.2.3 Composition of Analyzer

The UPT immunoassay analyzer mainly consists of four key units: photoelectric measurement system, control system, algorithm module and mechanical structure (Fig. 5.8).

To obtain high detection sensitivity and good linear response, the function and design requirements of each unit of the analyzer are as follows:

(1) The photoelectric measurement system has the following functions: (a) can generate a uniform illumination light spot of a specific wavelength, power density and size, (b) can efficiently collect the UCL signal within the measured wavelength range of the UCP particles on the LF strip with low noise, and (c) can sensitively and efficiently convert the collected UCL signal into an electrical signal.
(2) The control system is used to drive the analyzer devices, acquire the output signal of the photoelectric measurement system, and provide a user-friendly interface.
(3) The algorithm module is used to make the devices work in a reasonable way, process and analyze the output signal of the photoelectric measurement system.
(4) The mechanical structure supports and protects the analyzer components, and enable the analyzer simple and reliable.

5.3 Key Unit Techniques of UPT Immunoassay Analyzer

5.3.1 Photoelectric Measurement System

The photoelectric measurement system of a UPT immunoassay analyzer mainly includes an excitation light source, an optical system consisting of an illumination unit of the excitation light (hereafter called an illumination unit) and a UCL receiving unit (hereafter called a light receiving unit), a photoelectric converter to convert a UCL signal into an electrical signal and a signal processing circuit. Among them, the design of the optical system is key to the analyzer development.

UPT immunoassay analyzers generally are divided into two types, i.e., scanning type and imaging type (Fig. 5.9; Table 5.2). A scanning analyzer obtains the distribution of UCP particles on an LF strip by one-dimensionally (1-D) confocal scanning by the focal line of the focused excitation light, while an imaging analyzer uses a

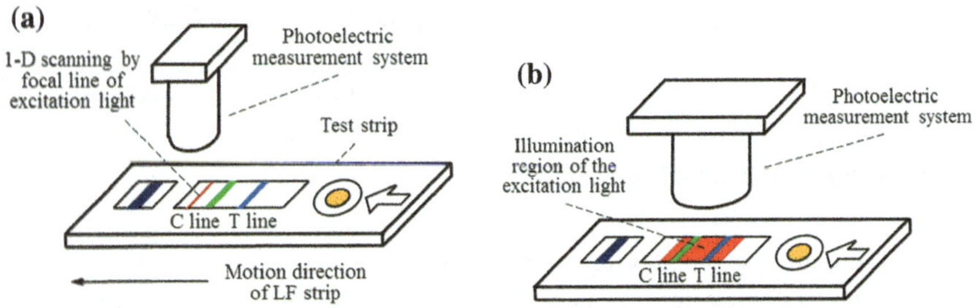

Fig. 5.9 Diagram of scanning (**a**) and imaging (**b**) technical schemes of UPT immunoassay analyzers

Table 5.2 Comparison of performance characteristics between scanning and imaging analyzers

Key unit		Scanning style	Imaging style
Photoelectric measurement system	Optical system	1-D confocal scanning by focal line is adopted, illumination region area is smaller, S/N ratio is high	Illumination region area is larger, uniformity requirement is higher, imaging quality requirement is higher
	Light source	Optical power requirement is lower	Optical power requirement is higher
	Photoelectric convertor	Use single-point photoelectric detector such as photomultiplier tubes or photodiode	Use image sensor
Mechanical structure		Need small and precise 1-D scanning stage	Do not need motion mechanism
Control system		Drive scanning stage to work	Drive image sensor to work
Algorithm module		Motion control, signal acquisition, data processing	Signal acquisition, data processing
Measurement time		Longer	Shorter

uniform excitation light spot to cover the test line and the control line and obtains the distribution of UCP particles on an LF strip with imaging optical system.

This chapter discusses the design of scanning analyzers. Imaging analyzers will be introduced in Sect. 5.4.3.

5.3.1.1 Optical System

1. Wavelengths of illumination light and signal light

The wavelengths of the illumination light and signal light of the optical system need to be determined first for the selection of the light source and the photoelectric converter.

Let us use the UCP of $NaYF_4$: Yb^{3+}, Er^{3+} as an example to describe the selection of wavelengths. The luminescence efficiency of $NaYF_4$: Yb^{3+}, Er^{3+} is highest under 980-nm infrared light excitation (Menyuk et al. 1972; Zhou et al. 2003). The emission spectrum of the UCP particles illuminated by the light at this wavelength is shown in Fig. 5.10, the emission primary peak wavelength is 541.5 nm, and the secondary peak wavelengths are 517 nm and 670 nm. The bandwidths of these emission peaks are all relatively narrow, and the shifts between these emission peaks and the excitation light wavelength are large. The curve is measured by an SPR-920D spectral radiometer (Hangzhou Zheda Sanse Instrument Co. Ltd., Hangzhou, China).

According to relevant research, the n values of the UCP particles corresponding to the UCL intensity within the above emission peaks and the excitation light power are all approximately equal to 2 (Suyver et al. 2005a, b, c, 2006; Zhao et al. 1999, 2000, 2005a). In addition, when the excitation light intensity increases, the proportion of the UCL intensity of the particles within the secondary peak at 670 nm increases significantly.

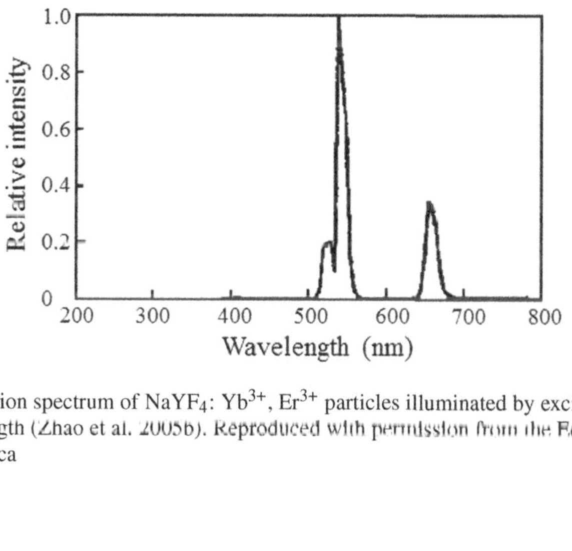

Fig. 5.10 Emission spectrum of $NaYF_4$: Yb^{3+}, Er^{3+} particles illuminated by excitation light with 980 nm wavelength (Zhao et al. 2005b). Reproduced with permission from the Editorial Office of Acta Optica Sinica

Therefore, for the $NaYF_4$: Yb^{3+}, Er^{3+} particles, the illumination light wavelength of the optical system is selected as 980 nm, and the signal light bandwidth should be between 500 and 700 nm, which includes the primary peak wavelength of 541.5 nm.

2. Structure of optical system

Two structures of co-axis and off-axis can be adopted in scanning and imaging analyzers.

In the co-axis design, the illumination unit and the light receiving unit may complete the two functions of excitation light focusing and UCL collection and collimation together by using a dichroic mirror that reflects the excitation light and transmits the UCL. This design has a compact structure and is beneficial to obtain a larger collection aperture angle.

In the off-axis design, usually, the optical axis of the light receiving unit is perpendicular to the surface of the LF strip, and has an inclined angle of 45° with the optical axis of the illumination unit.

5.3.1.2 Excitation Light Source and Illumination Unit

The light beam emitted from the excitation light source is transformed by the illumination unit into the required illumination light spot on the surface of the LF strip.

The selection of the excitation light source profits from the development of laser technology. The semiconductor laser, also known as the laser diode (LD), has the advantages of small size, light weight, high electro-optic conversion efficiency, long working life, low cost and the ability to be directly modulated. The main characteristic of the LD is that its output power and wavelength may vary with changes in temperature and current. Therefore, a temperature control device and constant power driving circuit are often added to this kind of laser to improve its stability to meet application requirements.

Therefore, a laser diode module (hereafter called an LD module) that is composed of a 980-nm wavelength LD and an aspheric lens with a large numerical aperture can be used as the excitation light source. This module can emit a collimated light beam with an approximatively rectangular cross section.

Because the actual distribution of UCP particles is not uniform in the test line and control line, to obtain the UCL signal of all UCP particles, the length of the focal line in the scanning analyzer needs to match the width of the result-scanning window of the LF strip, and the length and width of the rectangular illumination light spot in the imaging analyzer needs to match the length and width of the window (Fig. 5.9) (Zhao et al. 2006).

A cylindrical lens can be used in the illumination units of the two types of analyzers. The lens can focus the light beam emitted from the LD module into the focal line in a scanning analyzer or cause the beam to diverge one-dimensionally to form a rectangular light spot to cover the test line and control line of an LF strip in an imaging analyzer.

5.3.1.3 Light Receiving Unit and Photoelectric Converter

The UCL of UCP particles is weak. The quantity of UCP particles in the illumination region is less, and their UCL signal is lower. However, the power of the excitation light is much higher than that of the UCL, and stray light results in the light receiving unit due to the scattering of the excitation light on the LF strip surface. Therefore, on the one hand, the light receiving unit needs a large numerical aperture to the LF strip surface to improve the collection efficiency; on the other hand, it needs to have a high transmittance within the measured light bandwidth and a high suppression rate to stray light.

The confocal method by focal line is adopted in the optical system of a scanning analyzer to reduce the stray light from outside of the focal line, which improves the S/N ratio. The size of the slit field stop on the image surface of the receiving unit should only make the UCL of the UCP particles fall within the focal line pass through after being focused, and then reach the photoelectric converter.

The photoelectric converter needs to have high spectral responsivity or quantum efficiency, a high S/N ratio and low noise within the measured UCL bandwidth. The commonly used single-point photodetectors are mainly photomultiplier tubes (PMTs) and photodiodes (PDs), and the common image sensors are mainly CCD and CMOS image sensors. A PMT can convert an optical signal into an electrical signal and carry out secondary electron multiplication. It has the advantages of high sensitivity, a low dark current, short response time and good linearity. The cut-off wavelengths of some types of PMTs are less than 800 nm, which is very helpful in removing the influence of residual excitation light.

5.3.2 *Control System*

A UPT immunoassay analyzer should meet the requirements of automation, small size, multifunction, high speed, being powered by a built-in battery and low power consumption. Therefore, it needs to use an embedded control system. The embedded system (ES) is a special computer system including two parts of software and hardware. The system controls and coordinates each unit and the subsystems in the analyzer, and implements condition monitoring, signal acquisition, result output, human computer interaction and other special functions.

5.3.3 *Algorithm Module*

The algorithm module of a UPT immunoassay analyzer mainly has the functions of signal acquisition, data processing and function line searching, and should eliminate or reduce the following interference factors: (1) the noises generated by the collection of the UCL, photoelectric signal conversion and electric signal processing; (2) the background noise caused by the UCP particles nonspecifically stranded on the LF strip; and (3) the positioning errors of the function lines introduced in the processing, manufacture and loading of an LF strip.

The algorithm module is integrated into the software program and works via the control system.

5.3.4 *Mechanical Structure*

The mechanical structure of a UPT immunoassay analyzer need to guarantee the design requirements of the layout and each unit of the analyzer. It mainly includes the mechanical structure of the optical system, scanning stage and analyzer layout.

The mechanical structure of the optical system must ensure the relative space positioning of the various optical elements through rigid or elastic connection, fixation and adjustment between the optical elements and the mechanical parts to satisfy the optical system requirements of extinction, anti-vibration, thermal dissipation, dustproofing and other aspects. In order to facilitate the assembly of optical system and the integration and maintenance of analyzers, the optical system can be designed as an independent module.

In a scanning analyzer, an LF strip is loaded by the scanning stage and driven by the control system. The scanning stage needs to ensure design requirements such as scanning range, resolution, speed and positioning accuracy.

The analyzer layout and structure needs to meet the requirements including shading, heat-dissipation, electromagnetic compatibility, insertion and removal of LF strips and maintainability.

5.3.5 *Instrument Calibration*

To realize accurate quantitative detection, an UPT immunoassay analyzer needs to be calibrated for obtaining the response characteristic curve, so that the specifications of the analyzer can be evaluated. Meanwhile, the response characteristic curve is also a necessary condition for the actual use of the analyzer.

5.3.5.1 Repeatability, Stability and Reproducibility

The UPT biosensor consists of the UPT LF strip and the UPT immunoassay analyzer. The reliability of the UPT biosensor must be evaluated from the two aspects of the LF strip and the analyzer respectively, and usually is described by three specifications: repeatability, stability and reproducibility.

Repeatability refers to the consistency of repeated measurement results of the same LF strip in a short time using one analyzer. The performance can be evaluated by the relative standard deviation (*RSD*) or the coefficient of variation (*CV*) of the measurement results, which is equal to the ratio of the standard deviation (*SD*) and the mean:

$$CV = (SD/mean) \times 100\%. \tag{5.2}$$

A smaller *CV* value reflects better repeatability (Miao 2005).

Stability refers to the extent to which the performance of the analyzer remains constant with time and is usually the consistency of the long-term performance of the analyzer. The performance can be evaluated by the *CV* value of multiple measurement results to the same LF strip with a longer time interval and using one analyzer.

Reproducibility refers to the extent of the similarity of measurement results to the same LF strip by using multiple analyzers of the same type. The performance usually is expressed by the *CV* value of the measurement results to the same LF strip by using these analyzers, and it can be used to assess the maturity of the batch manufacturing process of the analyzers. In general, a *CV* value less than 10% is acceptable, especially in the trail production phase.

5.3.5.2 Response Characteristic Curve

By preparing a series of standard LF strips of known target analyte concentrations, a group of signals (S_T/S_C values) corresponding to the concentrations can be obtained by an analyzer. Based on a certain mathematical model, the response characteristic curve of the UPT biosensor can be fitted according to the results and used for quantitative detection.

In the curve fitting, the linear fitting equation obtained by the least square method has only two parameters and can describe most of the rules of biological reactions.

The response characteristic curve of a UPT biosensor can be divided into three segments (Fig. 5.11). The segment A indicates that the target analyte concentration is below the lower limit of detection C_{min} of the UPT biosensor, and the segment C indicates that the concentration is higher than the upper limit of detection C_{max}. In these two segments, the detection results cannot accurately correspond to the actual

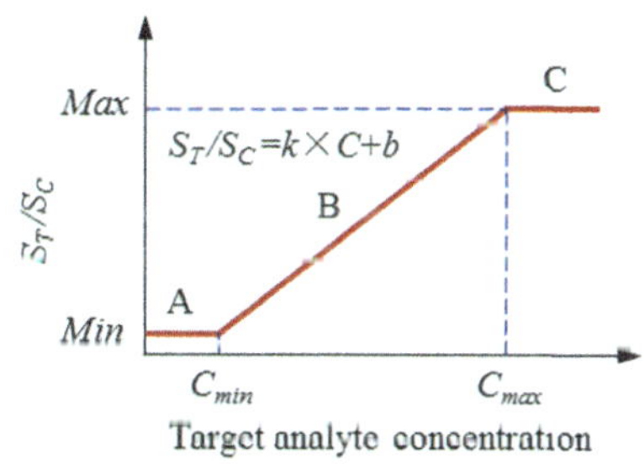

Fig. 5.11 Schematic of linearly fitted response characteristic curve

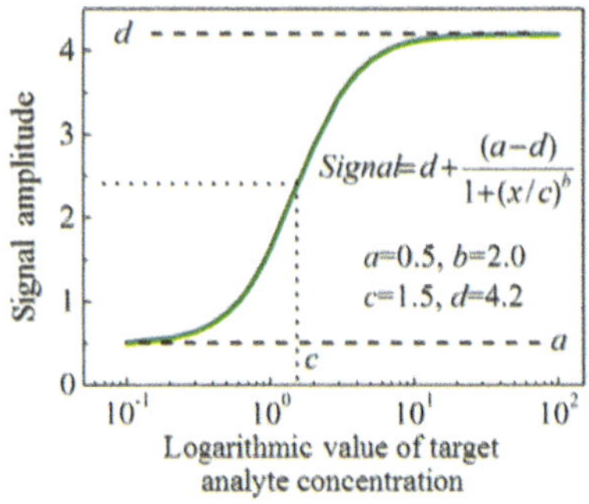

Fig. 5.12 Response characteristic curve based on four-parameter logistic model

target analyte concentrations. The segment B, in which the target analyte concentration ranges from C_{min} to C_{max}, is the linear range of the UPT biosensor. Through the concentrations of the standard samples in the segment B and the corresponding detection results, the linear equation $S_T/S_C = k \times C + b$ can be fitted out. In this equation, k is the slope of the fitted line, which is the sensitivity of the sensor in this linear range, i.e., the parameter that evaluates the capability of the sensor to detect trace matter, and b is the intercept of the line.

The four-parameter logistic model (hereafter called the 4PL model) is an effective and universal model, and its general form is

$$Signal = d + \frac{(a - d)}{1 + (x/c)^b}, \tag{5.3}$$

where x is the target analyte concentration, a is the detection result of the analyzer when there is no target analyte in the sample, b is the slope factor indicating the variation of the detection result when the target analyte concentration (or dose) increases, c is the target analyte concentration (or dose) when the detection result is the median between a and d, and d is the detection result of the analyzer when the target analyte in the sample is excessive (Fig. 5.12) (Huang et al. 2009).

When the immunoreaction mode, the target analyte type or the LF strip production batch changes, and after the analyzer operates for a period of time or is adjusted, the response characteristic curve of the UPT biosensor needs to be re-calibrated.

5.4 Development of UPT Immunoassay Analyzer

Based on the scheme of the scanning measurement technique, five models of four generations of UPT immunoassay analyzers were developed (Table 5.3). The first three generations, four models in total, are single-channel UPT immunoassay analyzers, and the fourth generation is a multichannel UPT immunoassay analyzer (UPT-M

Table 5.3 Summary of UPT immunoassay analyzers

Model	Optical system	Control system	Power supply	Measurement time	Application
UPT-1	Co-axis structure	Based on 3.5″ SBC	External power supply	90 s	Prototype
UPT-2	Off-axis structure	Based on PC/104 SBC	Built-in battery	30 s	Small batch application
UPT-3		Based on SCM	Built-in battery	30 s	Small batch application
UPT-3A		Based on ARM micro-controller	Built-in battery	20 s	Product
UPT-M		Based on PC/104 SBC	External power supply	2 min	Prototype
Imaging		Based on SCM		10 s	Experimental system

analyzer). Furthermore, we developed an imaging detection system for the LF strip using a linear image sensor.

5.4.1 Development of Single-Channel UPT Immunoassay Analyzer

5.4.1.1 UPT-1 Analyzer

The UPT-1 analyzer is designed based on one-dimensionally confocal scanning by the focal line, and the working principle is shown in Fig. 5.13 (Zhao et al. 2005b, 2006; Lu et al. 2006).

1. Photoelectric measurement system

The optical system of the UPT-1 analyzer adopts a co-axis structure, i.e., the illumination unit and the light receiving unit share a focusing/collimating lens through the dichroic mirror.

The collimated light beam with an approximatively rectangular cross section emitted from the LD module is one-dimensionally focused by the convex-plane cylindrical lens, and is then reflected by the dichroic mirror and focused by the focusing/collimating lens into the focal line on the surface of the LF strip. The dichroic mirror reflects the excitation light and transmits the UCL, and the angle between the reflecting surface and the incident light beam is 45°.

The UCL of the UCP particles within the illumination region is collected and collimated by the focusing/collimating lens, then passes the dichroic mirror and the optical filter used to filter out the stray light, and then is focused on a slit field stop by

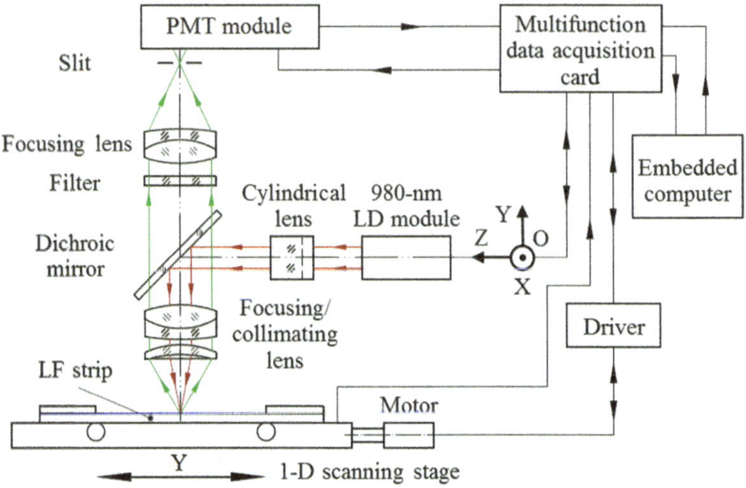

Fig. 5.13 Schematic of working principle of UPT-1 analyzer. Modified from Zhao et al. (2005b)

the focusing lens. The optical signal that has passed through the slit incidents on the photosensitive surface of the photoelectric converter. The converted electrical signal is then amplified and transmitted to the control system.

(1) Excitation light source and illumination unit

The LD module with a 980-nm wavelength was selected as the light source. The output power of the LD module was 30 mW, and the emission spectrum is shown in Fig. 5.14 (measured by a SPEX 1702/04 spectrometer). The width *W* along the X axis and height *H* along the Y axis of the rectangular cross section of the collimated light beam emitted from the LD module were 4 mm and 2 mm, respectively.

The dichroic mirror in the illumination unit needs to have high reflectivity for the excitation light and high transmittance of the UCL. The transmittance curve of the dichroic mirror used when the angle between the reflecting surface and the incident light beam is 45° is shown in Fig. 5.15 (measured by a PerkinElmer Lambda 900

Fig. 5.14 Emission spectrum of LD module (Lu et al. 2006). Reproduced with permission from the Editorial Department of Acta Photonica Sinica

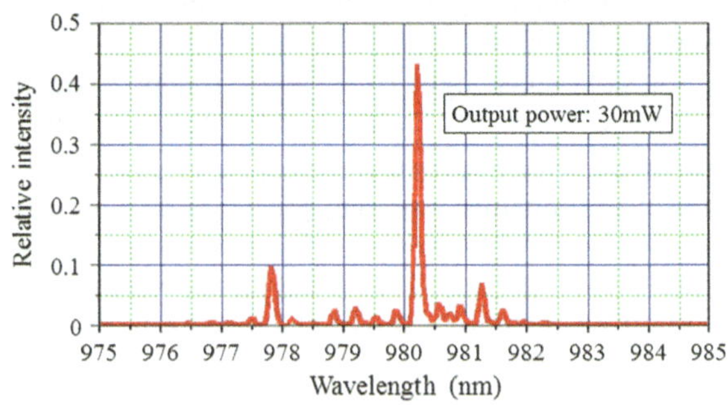

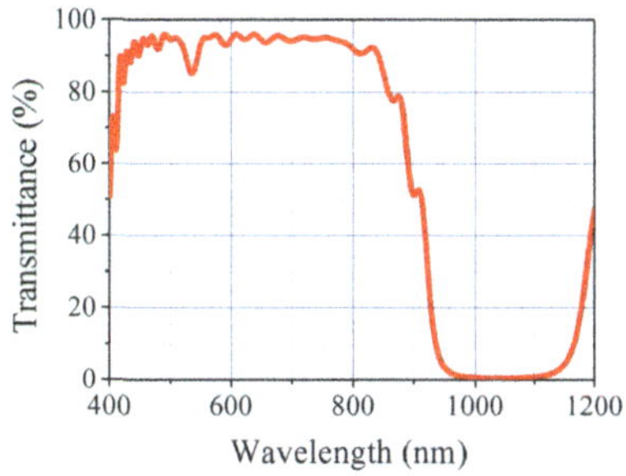

Fig. 5.15 Spectrum transmittance curve of dichroic mirror when angle between reflecting surface and incident light beam is 45° (Zhao et al. 2005b). Reproduced with permission from the Editorial Office of Acta Optica Sinica

spectrophotometer). The transmittance at 980 nm was about 0.88%. If the loss of light energy is neglected, the reflectivity is about 99.12%. The transmittance ranging from 500 to 700 nm was from 85 to 95%.

The focusing/collimating lens shared by the two units consisted of an achromatic doublet lens and an aplanatic lens. The numerical aperture (NA) of the focusing/collimating lens was 0.386, and the focal length $f_d{}'$ was 17.659 mm. The focal length of the cylindrical lens was 72.2 mm.

The light beam emitted from the LD module remained collimated in the YOZ plane after passing through the cylindrical lens, then was focused on the LF strip surface by the focusing/collimating lens. The light beam was in the defocused position along the X axis on the LF strip surface.

The focal depth of the collimated light beam focused by the focusing/collimating lens is

$$\delta f' = \lambda/[H/(2f')]^2. \quad (5.4)$$

The height H of the collimated light beam was 2 mm. Therefore, $\delta f'$ was calculated to be about 0.3 mm by the above equation.

According to the Fraunhofer diffraction equation of rectangular apertures, the half width of the focal spot of the collimated beam focused by the focusing/collimating lens is

$$\rho = \lambda/(H/f'). \quad (5.5)$$

Because the actual incident beam has a divergence angle (half angle), the focal line on the focal plane of the focusing/collimating lens widens. Therefore, the total width of the focal line is

$$w = 2[f'\theta + \lambda/(H/f')]. \quad (5.6)$$

The divergence angle of the light beam emitted from the LD module was about 0.6 mrad. From this, the width w of the focal line was calculated to be 38.5 μm. Meanwhile, the length of the focal line of the excitation light (along the X axis) was calculated to be 0.852 mm by optical design software.

(2) Light receiving unit

The light receiving unit consists of the focusing/collimating lens, a filter, a focusing lens and a slit field stop.

The filter in the light receiving unit is used to filter the stray light outside the signal light bandwidth, suppress the background noise of the optical system, and improve the S/N ratio. The transmittance curve of the filter measured by a PerkinElmer Lambda 900 spectrophotometer is shown in Fig. 5.16. The transmittance was higher than 95% at the 541.5-nm wavelength, and was less than 0.01% at the 980-nm excitation wavelength.

The magnification of the light receiving unit is $\beta = f_d'/f'$, where f_d' is the focal length of the focusing lens. The focusing lens is the same as the achromatic doublet lens in the focusing/collimating lens, and f_d' was 25.243 mm. Thus, β was calculated to be $-1.429\times$, and the size of the image on the back focal plane of the focusing lens of the excitation light focal line on the LF strip was about 1.218 mm × 0.055 mm.

The spot diagram calculated by optical design software is shown in Fig. 5.17, the size of the dispersion circle of the center field, the diameter of which was about ϕ0.186 mm, was close to that of the off-axis field. In the actual design, the size of the slit field stop, for its assembly and adjustment, was set to 2 mm × 0.6 mm.

(3) Photoelectric converter

A H5784-01 PMT module (Hamamatsu Photonics KK) was used as the photoelectric converter. The module includes a metal package PMT, a low power consumption high-voltage power supply and a low-noise pre-amplifier. The effective area's diameter of the photoelectric surface of the PMT was ϕ8 mm. The spectral response range of the module was 300–850 nm, the peak wavelength was 400 nm, and there was no response to the excitation light at the 980-nm wavelength. The feedback resistance of the preamplifier was 1 MΩ, the current-to-voltage conversion factor was 1 V/μA, and the frequency bandwidth was 20 kHz (Hamamatsu Photonics 2008).

2. Control system

The hardware of the UPT-1 analyzer control system is mainly composed of a 3.5″ embedded SBC and a multifunction data acquisition (DAQ) card connected via a PC/104 interface (Fig. 5.18).

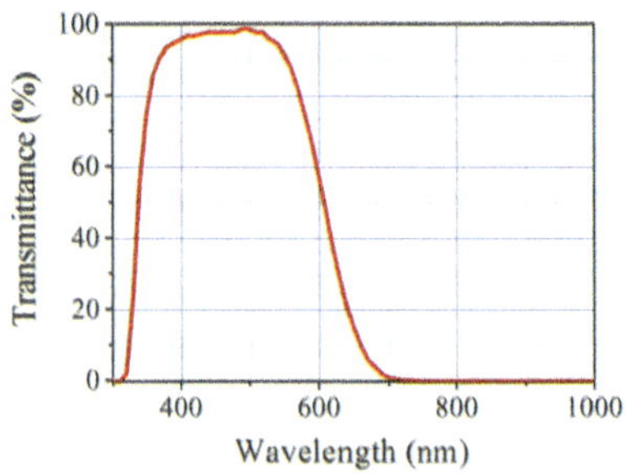

Fig. 5.16 Spectrum transmittance curve of filter

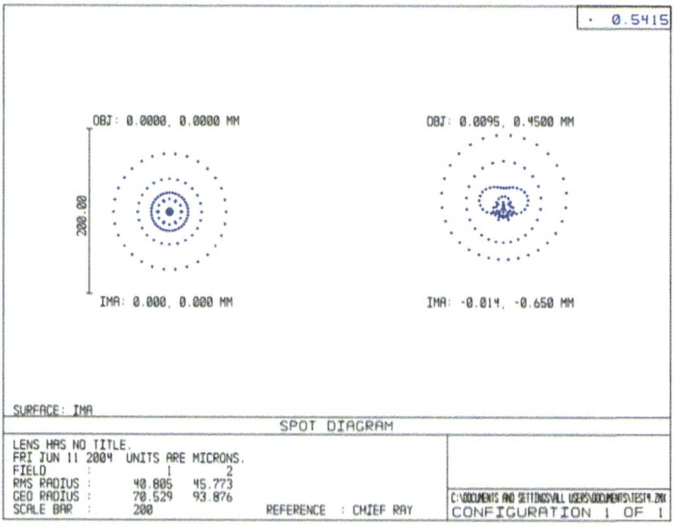

Fig. 5.17 Spot diagram of different view fields of light receiving unit

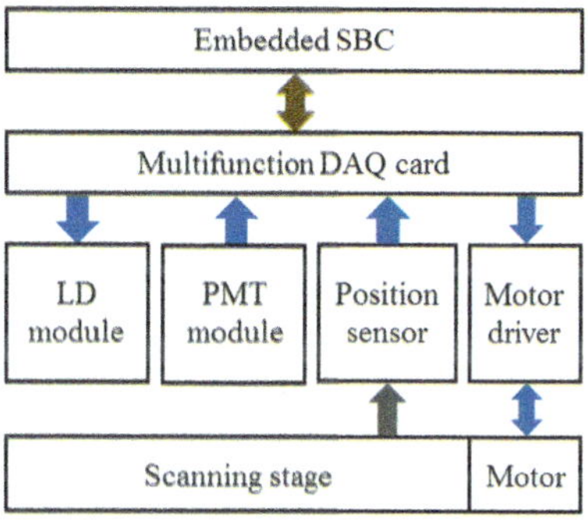

Fig. 5.18 Hardware composition of UPT-1 analyzer control system

The functions of the control system include:

(1) power-on initialization and the condition monitoring of the analyzer, including parameter setting, status inquiry and feedback,
(2) analyzer automatic measurement control, including the motion control of the scanning stage, acquisition of the signal output from the photoelectric measurement system and result output,
(3) a user-friendly interface to allow an operator to obtain information instantly and clearly, to input or select instructions accurately, and to correct or verify the settings or detection results of the analyzer, and
(4) functions needed for the test, calibration, examination, maintenance and other operations to the analyzer.

The detailed program flow of the analyzer automatic measurement control is shown in Fig. 5.19. Among them, the control system, at the beginning of the measurement, determines whether to carry out the reset operation of the scanning stage according to the judgement result of the LF strip position by the status of the two Hall proximity switches corresponding to the start and the end points of the result-scanning window of the LF strip (The directions of the reset and forward movement of the scanning stage are controlled by the direction signal input into the motor driver from the control system). When the LD stabilizes, the control system acquires the voltage signals output from the PMT module according to the signal acquisition algorithm, performs data processing, stores the results into memory, and then sends the required number of pulses to the motor driver, which moves the LF strip forward to the next measurement position.

When the measurement ends, a raw data curve is drawn in the dialog box on the screen of the analyzer.

3. Algorithm module

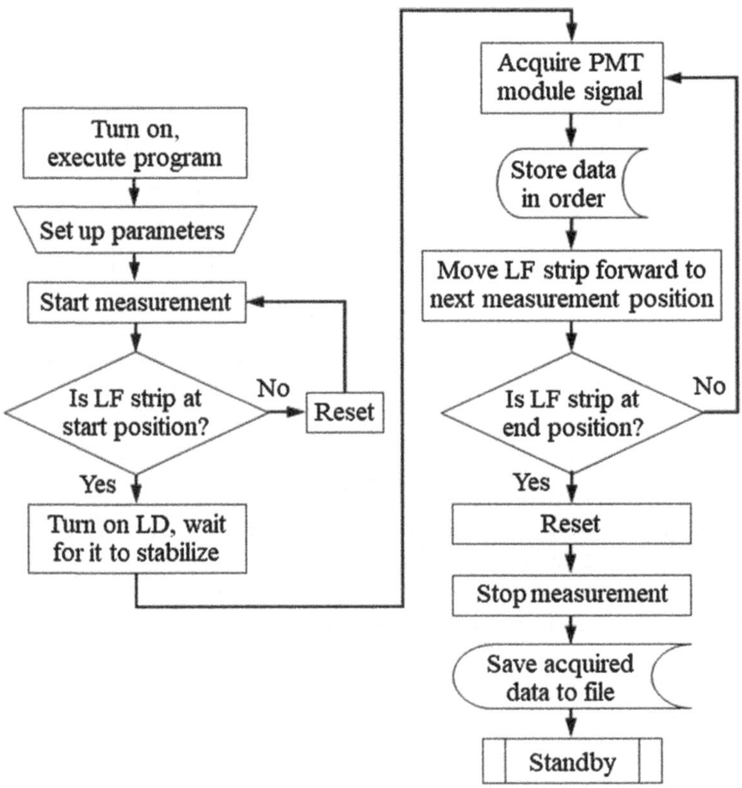

Fig. 5.19 Automatic measurement control flow chart

The main function of the algorithm module of the UPT-1 analyzer is data processing, including three steps: data curve smoothing and denoising, function line search and result calculation. The detailed flow is shown in Fig. 5.20. Along with S_T/S_C being used as the detection result, the first two steps are implemented based on their respective algorithms.

(1) Smoothing and denoising algorithm

The UCL signal distribution measured by the UPT-1 analyzer is shown as the violet curve 1 in Fig. 5.21a, which is shifted upward by 1 V artificially for the convenience of comparison. The curve shows many rapid fluctuations and the background noise coming from the UCP particles nonspecifically stranded. Therefore, a five-point value smoothing for reducing fluctuations and a baseline translation processing for denoising were performed on the data in the measurement region, assuming the background noise distribution was linear.

The smoothed result of the raw data curve is shown as the red curve 2 in Fig. 5.21a. Then the denoising is shown in Fig. 5.21b: the minimum points searched near both boundaries of the test line in the curve 2, which was artificially shifted upward by 1 V, were selected as the start and end points of the background noise baseline of the test line and the noise value of each point was calculated. Similarly, the background noise baseline of the control line and the noise value of each point can be obtained. Thus, the blue curve 3 was obtained by removing noise from the curve 2. By using this relatively simple data processing algorithm, the interference of the fluctuations and the background noise were reduced to a certain extent.

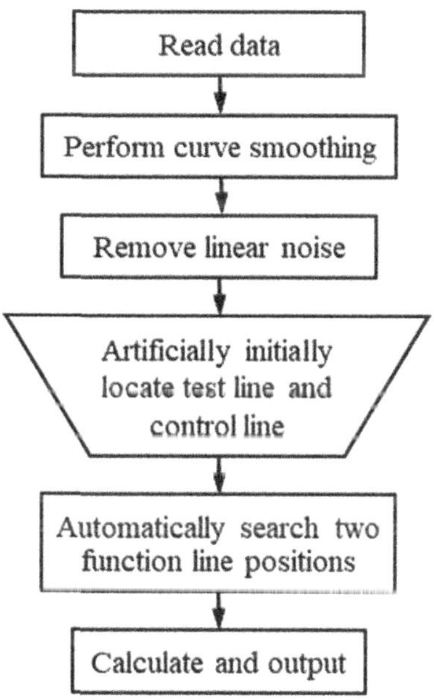

Fig. 5.20 Data processing flow chart. Modified from Lu et al. (2006)

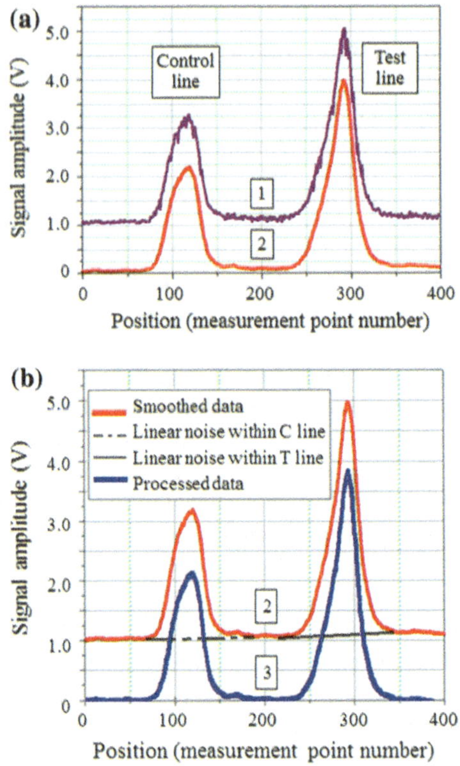

Fig. 5.21 Schematic of smoothing and denoising algorithm

(2) Function line semi-automatic search algorithm

The function lines of an LF strip have positioning errors that do not allow the analyzer to directly and accurately calculate the signal S_T of the test line and S_C of the control line. Therefore, on the basis of the artificial initial location of the function lines, an average calculation of the adjacent multiple points is made to the smoothed and denoised curve according to the widths of the function lines, then the maximum values automatically searched in the two function lines are taken as the signals of S_T and S_C, and the coordinates of the values are set as the function line locations.

4. Calibration

First, the detection repeatability and the detection stability of the UPT-1 analyzer were evaluated using a double-antigen sandwich LF strip added with the rabbit anti-plague Immunoglobulin G (IgG) sample. In addition, the LF strip was repeatedly inserted into the analyzer and measured to evaluate the validity of the function line search algorithm of the analyzer to LF strips. Each of the above three specifications was evaluated by the coefficients of variation (CV) of 12 measurement results (Table 5.4). The CV values of S_T/S_C in the evaluation results were all less than 5%, which indicates that the analyzer had good detection repeatability and stability, and was

Table 5.4 Evaluation results of detection reliability of UPT-1 biosensor

Evaluation content	*CV* values of measurement results		
	S_T (%)	S_C (%)	S_T/S_C (%)
Repeatability	6.26	6.285	1.224
Stability	7.408	6.507	4.693
Validity of function line search algorithm	2.879	1.499	3.0

less affected by the positioning error of the LF strip, and therefore had reliable performance.

The detection results of a series of 20 rabbit anti-plague IgG standard samples with different concentrations using double-antigen sandwich LF strips and the UPT-1 analyzer are shown in Table 5.5.

Taking the linear fitting to the detection results in the concentration range of 200–6000 ng/ml, the linear equation obtained is $Y = -0.15708 + 0.000618684 \times$

Table 5.5 Detection results of rabbit anti-plague IgG standard samples with different concentrations using a UPT-1 analyzer

LF strip number	Concentration (ng/ml)	S_T (V)	S_C (V)	S_T/S_C
1	0	0.08022	2.53308	0.03167
2	200	0.21162	1.96276	0.10782
3	400	0.32219	1.41627	0.22749
4	600	0.33713	1.44870	0.23271
5	800	0.44608	1.18135	0.37760
6	1000	0.55548	1.05232	0.52786
7	1500	0.38261	0.88773	0.43100
8	2000	0.90733	0.83361	1.08843
9	2500	0.94500	0.78799	1.19926
10	3000	1.49448	0.79857	1.87144
11	3500	1.05374	0.67269	1.56645
12	4000	1.51084	0.65686	2.30010
13	4500	2.12359	0.71419	2.97343
14	5000	1.84586	0.62598	2.94876
15	5500	1.88203	0.56959	3.30419
16	6000	2.29032	0.64625	3.54402
17	6500	1.80184	0.63839	2.82248
18	7000	2.56942	0.48019	5.35083
19	7500	2.35049	0.64757	3.62971
20	8000	2.47013	0.41861	5.90079

Detection result of S_T/S_C is from Zhao et al. (2005b)

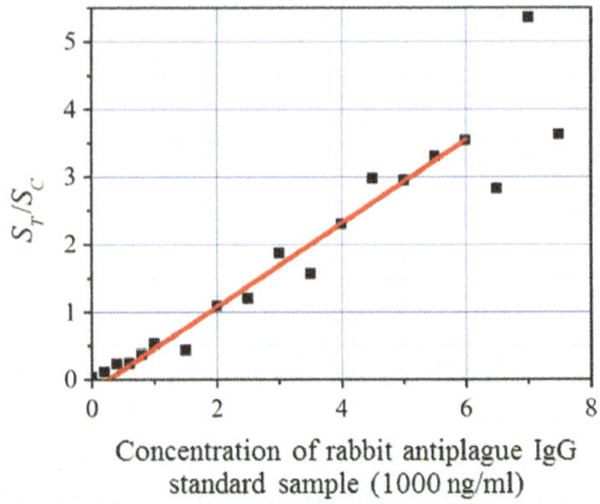

Fig. 5.22 Detection results of rabbit antiplague IgG standard samples with different concentrations using UPT-1 analyzer and fitted response characteristic curve. Modified from Zhao et al. (2005b)

Fig. 5.23 UPT-1 analyzer (Zhao et al. 2005b). Reprinted with permission from the Editorial Office of Acta Optica Sinica

X, where X is the concentration of the sample, and Y is the S_T/S_C value (Fig. 5.22). The correlation coefficient r of the fitting equation is 0.98707, which indicates that the response linearity of the UPT biosensor in the concentration range was good.

A photograph of the developed UPT-1 analyzer is shown in Fig. 5.23.

5.4.1.2 UPT-2 Analyzer

The UPT-2 analyzer is a single-channel scanning UPT immunoassay analyzer that uses an optical system with an off-axis structure, and the optical system principle is shown in Fig. 5.24 (Huang et al. 2009).

1. Photoelectric measurement system

In the optical system of the UPT-2 analyzer, the angle between the optical axis of the illumination unit and that of the light receiving unit, which is perpendicular to the LF strip surface, is 45°.

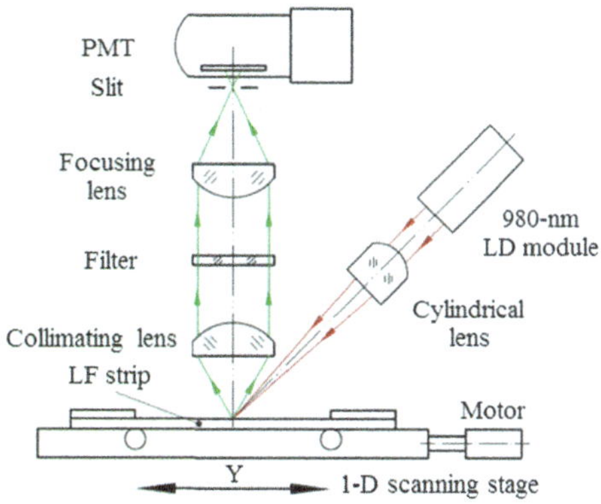

Fig. 5.24 Schematic of optical system principle of UPT-2 analyzer. Modified from Huang et al. (2009)

An LD module with a 980-nm wavelength was used as the excitation light source, and the output power of the module was 100 mW. A cylindrical lens with a 32.75-mm focal length was used in the illumination unit to one-dimensionally focus the approximately collimated light beam emitted from the LD module into a rectangular focal line of 3 mm × 50 μm on the LF strip surface.

The filter in the light receiving unit was the same as that in the UPT-1 analyzer, both the collimating lens and the focusing lens were a plano-convex lens with an NA of 0.36. The size of the rectangular slit behind the focusing lens was 3.5 mm × 0.6 mm.

2. Control system

The control system hardware of the UPT-2 analyzer uses a 7020B embedded SBC and an ADT652 multifunction data acquisition card which conform to the PC/104 standard and are manufactured by SBS Science & Technology Co., Ltd. The analyzer is powered by a built-in rechargeable lithium battery.

A small stepper motor is used as the actuator of the LF strip scanning stage in the analyzer, its motion control is implemented by the hardware and software of the control system and an efficient algorithm.

3. Algorithm module

A multi-thread software technique is used for motion control, and the code that drives the motor is located in an auxiliary thread to improve the efficiency of the CPU. The square wave output function and the timing function of the two timers in the data acquisition card are used to generate the sequential pulse signals needed to drive the motor. This method requires only initialization, and does not occupy CPU processing time. Therefore, it can save CPU resources and obtain a precisely timed pulse output to stabilize the motor movement and ensure an accurate scanning resolution.

The data processing algorithm comprises four parts (Fig. 5.25).

The principle of the algorithm is shown in Fig. 5.26. First, a Fourier transform is used on the raw one-dimensional distribution signal for the analysis of its spatial spectrum and the determination of its cut-off frequency, the signal is low-pass filtered

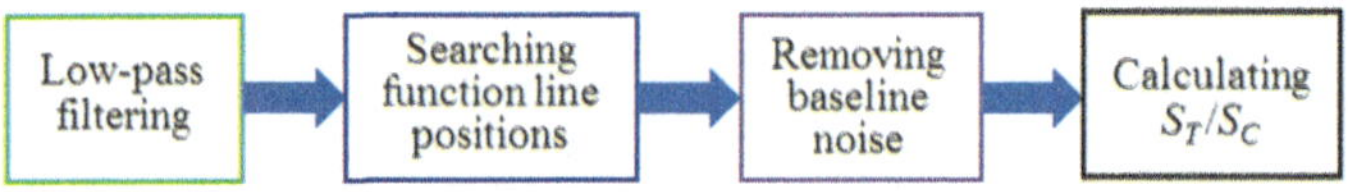

Fig. 5.25 Data processing algorithm flow chart

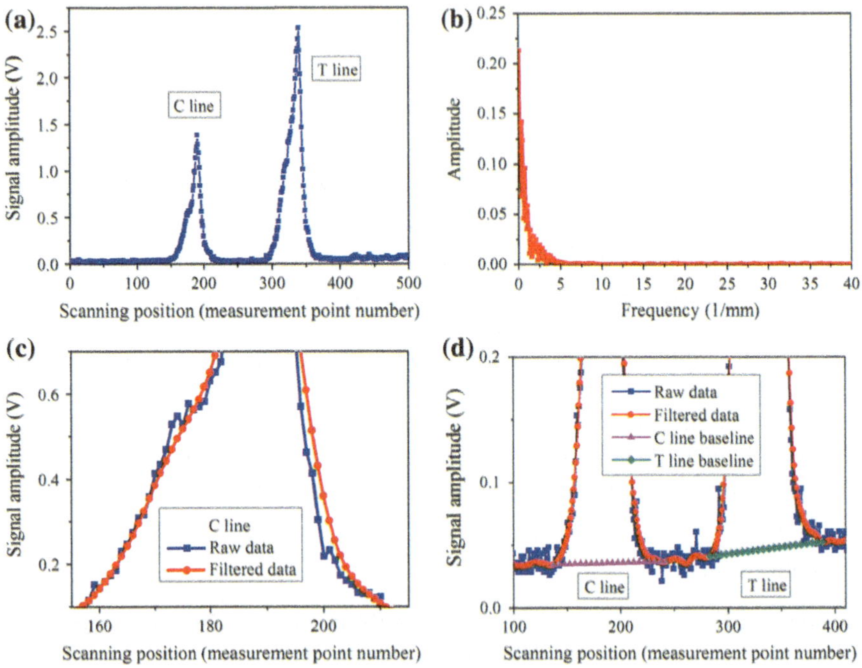

Fig. 5.26 Schematic of data processing algorithm (Huang et al. 2009). Reproduced with permission from IEEE

by using the Remez algorithm, and the resulting data are stored. Then, the first-order derivative of the data is solved, and the boundary positions of the test line and the control line are determined on the basis of the positive and negative variations of the derivative. The baseline equation of the noise is determined according to the boundary values of the two function lines, and the corresponding noise values are removed from the data after filtering. Finally, S_T and S_C are calculated by the summation of the data within the test line and the control line, and the target analyte concentration of the sample is obtained according to S_T/S_C and the response characteristic curve.

4. Calibration

First, the detection repeatability and the detection reproducibility of the UPT-2 analyzer were evaluated using a double-antibody sandwich LF strip that was added with a 10 ng/ml plague F1 antigen standard sample and on which the reaction was stable (Fig. 5.27).

The 20 repeated measurement results of the LF strip using one UPT-2 analyzer are shown in Fig. 5.27a. The *CV* value of 2.68% indicates that the analyzer had good detection repeatability. The measurement results of 12 analyzers of a small batch production are shown in Fig. 5.27b. The *CV* value of 2.9% indicates that the detection reproducibility of the analyzer batch was good, and the manufacturing process was relatively mature. Therefore, the performance of the UPT-2 analyzer was reliable.

The detection results of a series of plague F1 antigen standard samples with different concentrations using double-antibody sandwich LF strips and the UPT-2 analyzer are presented in Table 5.6. The result of 3.147 when the concentration was

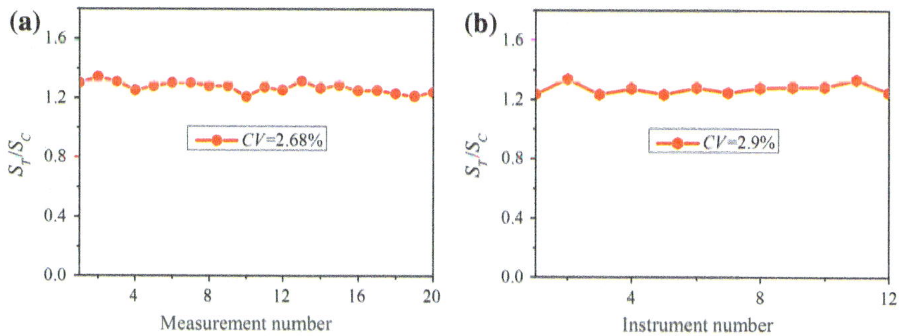

Fig. 5.27 Evaluation results of UPT-2 analyzer: detection repeatability (**a**) and reproducibility (**b**)

Table 5.6 Detection results of plague F1 antigen standard samples with different concentrations using a UPT-2 analyzer (Huang et al. 2009)

LF strip number	Sample concentration (ng/ml)	S_T/S_C
1	0	0.373
2	5	1.003
3	10	1.270
4	20	1.633
5	40	2.142
6	60	2.673
7	80	3.284
8	100	3.327
9	150	3.474
10	200	3.147

Reproduced with permission from IEEE

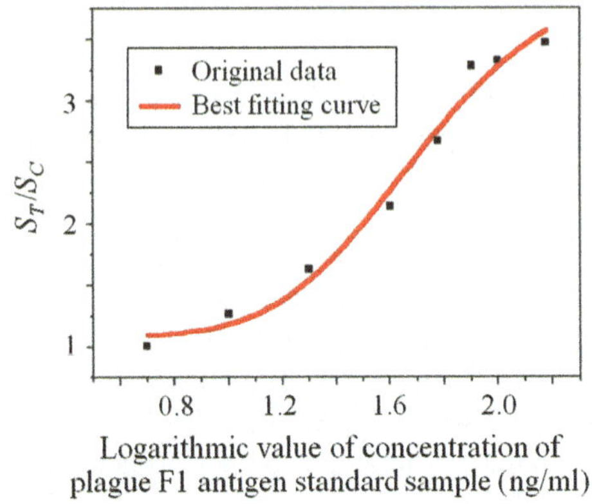

Fig. 5.28 Detection results of plague F1 antigen standard samples with different concentrations using UPT-2 analyzer and response characteristic curve fitted on basis of 4PL model (Huang et al. 2009). Reproduced with permission from IEEE

200 ng/ml was less than that of 3.474 when the concentration was 150 ng/ml. This phenomenon is called the Hook effect or the Prozone effect, and occurs when there are more plague F1 antigens in the sample. The binding between the antigens and antibodies in the test line could weaken the binding between the F1 antigen-UCP-antibody A conjugates and the antibodies in the test line, which ultimately decreases the number of UCP particles in the test line and causes the measurement value to decline.

The response characteristic curve fitted according to the detection results of the samples with concentrations from 5 to 150 ng/ml using the 4PL model is shown in Fig. 5.28. The equation is

$$Y = 4.191 + \frac{(1.084 - 4.191)}{1 + \left(X/1.731\right)^{6.095}},$$

where X is the logarithmic value of the concentration, Y is the S_T/S_C value, and r^2 is 0.984, indicating that the response characteristic of the UPT biosensor in the concentration range was good.

A photograph of the developed UPT-2 analyzer is shown in Fig. 5.29.

5.4.1.3 UPT-3A Analyzer

The UPT-3A analyzer is designed for clinical diagnostic market. The analyzer uses an optical system with an off-axis structure and a control system based on the ARM microcontroller, and is powered by a built-in lithium battery.

RFID electronic tags integrated into the LF strips that are detected by the UPT-3A analyzer contain the product information and quantitative parameters. The analyzer

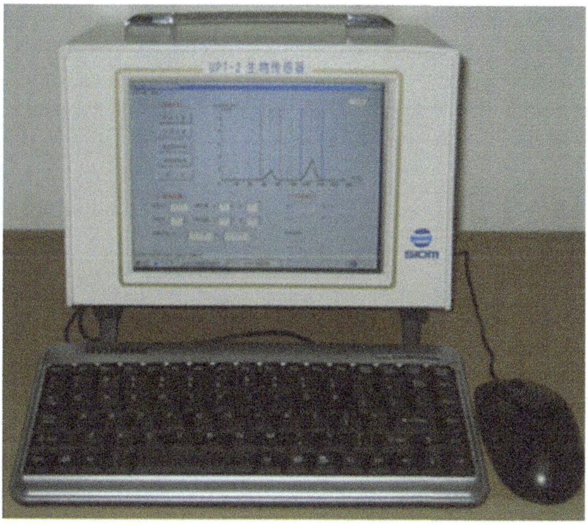

Fig. 5.29 UPT-2 analyzer

obtains these data through contactless access to the tag by using a built-in RFID read-write module.

1. Control system

The control system hardware of the UPT-3A analyzer contains a core STM32F103 MCU (STMicroelectronics). In addition to the motion control of the scanning stage, signal acquisition and data processing, the control system supports contactless access to the LF strip, keyboard operation, LCD display, data storage, record backup, the printing of results, a real-time clock, battery monitoring and PMT gain control (Fig. 5.30). The RS232 serial debugging interface, RFID read-write module, micro printer and other peripherals communicate with the STM32F103 MCU through a USART bus interface. The PMT gain control chip, nonvolatile ferroelectric RAM, SD card and other peripherals communicate with the STM32F103 MCU through an SPI interface.

The software architecture of the control system is shown in Fig. 5.31. When the analyzer is turned on, the control system initializes by reading the protection parameters and the user interface display, performs other preparatory tasks, and then accesses the function program. After detecting and updating the status information of each part of the analyzer, the control system scans the keyboard input instructions, and performs the requested operation.

2. Algorithm module

The data processing algorithm used in the UPT-3A analyzer first performs cross-correlation processing of the signals after smoothing, and then performs an adaptive function line boundary orientation (Xie et al. 2009).

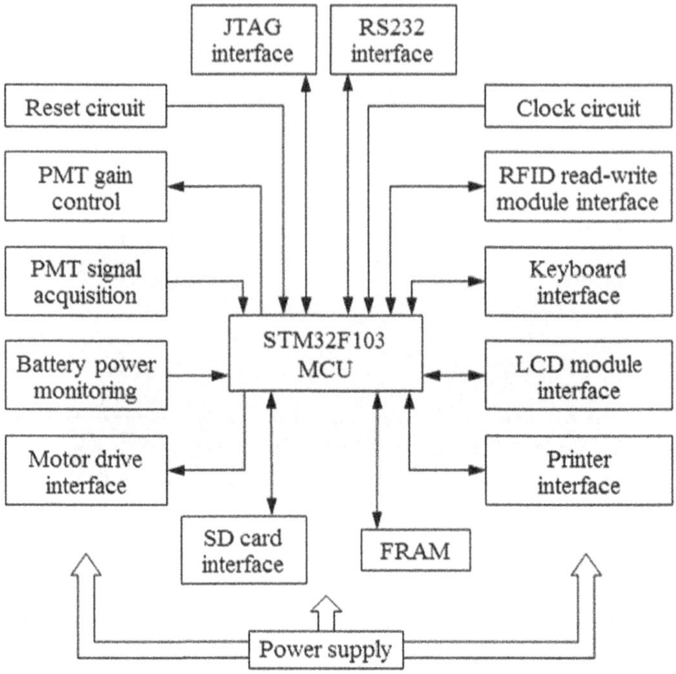

Fig. 5.30 Schematic of hardware architecture of UPT-3A analyzer control system

Fig. 5.31 Schematic of software architecture of UPT-3A analyzer control system

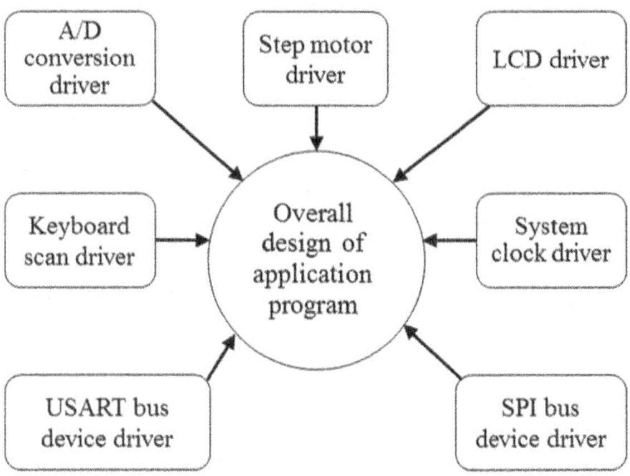

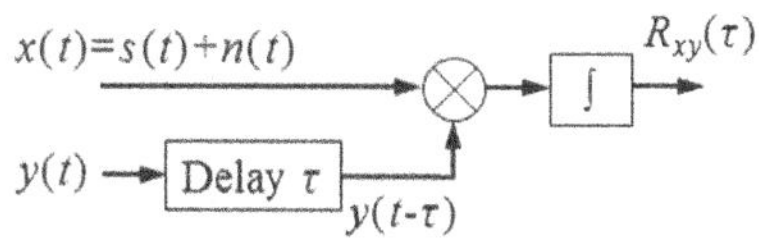

Fig. 5.32 Schematic of cross correlator (Xie et al. 2009)

The correlation function describes the relation or degree of similarity between two signals, and can be applied to the detection, recognition and extraction of the signals in random signal analysis and processing. The correlation function includes the auto-correlation function and the cross-correlation function (CCF); the CCF describes the association degree between two different random signals.

The principle of a CCF calculator (cross-correlator) for time-domain signals is shown in Fig. 5.32 (Ji and Zhang 2004). The input signal is $x(t) = s(t) + n(t)$, where $s(t)$ is the measured signal, $n(t)$ is the noise signal, and $y(t)$ is the input reference signal.

Suppose $y(t)$ is correlated to the measured signal $s(t)$, and is uncorrelated with the noise signal $n(t)$, the output of the cross-correlator is the CCF of $x(t)$ and $y(t)$ given by

$$R_{xy}(\tau) = \lim_{T\to\infty} \frac{1}{2T} \int_0^T [s(t) + n(t)]y(t - \tau)dt = R_{sy}(\tau) + R_{ny}(\tau). \tag{5.7}$$

Since $n(t)$ is uncorrelated with $y(t)$, $R_{ny}(\tau) = 0$,

$$R_{xy}(\tau) = R_{sy}(\tau). \tag{5.8}$$

Therefore, the purpose of the cross-correlator to suppress noise is achieved.

For a finite-length discrete sequence signal $x(n)$, where $n = 0 - (N - 1)$, the calculation formula of its CCF is

$$R_{xy}(m) = \frac{1}{N - m} \sum_{n=0}^{N-1-m} x(n)y(n + m), \quad m = -(N - 1) \sim (N - 1). \tag{5.9}$$

If a sample is detected by using a double-antigen sandwich LF strip, the binding of the UCP-antigen A-target analyte conjugates in the sample to the corresponding antigens within the width of the test line and that of the free UCP-antigen A conjugates to the antibodies within the width of the control line are both Gaussian random processes when the two types of conjugates flow together through the analytical membrane. Accordingly, the quantities of the UCP particles result within the two functional lines due to the specific binding both present Gaussian distributions with the widths of the function lines. The case of an indirect LF strip is similar. According to the linear correspondence of the signal conversion in the analyzer, the voltage signals acquired in the two functional lines follow the normal distribution law. Therefore, the Gaussian function can be selected as the reference signal, and its prototype algorithm is given as

$$f(x) = a \times e^{-(x-b)^2/c^2}, \tag{5.10}$$

where a is preferable to 1. The LF strip processing has become mature after the UPT-2 analyzer. The widths of the test line and the control line are both 1 mm, the distance between the centers of the two function lines is 3 mm, and the scanning range is 9 mm. Accordingly, when the scanning resolution is 50 μm, a total of 180 measurement points exist. The reference signal is symmetrically distributed by setting b as 90, which is half of the total points. According to the verification of the experimental simulation, the effective width of the reference signal was 20 when $c^2 = 16$, which meets the requirements of the cross-correlation processing of the measured signals.

The measured signal after the preliminary smoothing filter, i.e., the input signal, was cross-correlated with the Gaussian function reference signal, and the result is shown in Fig. 5.33. The result indicates that cross-correlation processing can effectively suppress the random noise of an input signal, and is conducive to further analysis and processing.

The signals, after being cross-correlationally processed, are mainly composed of background signals, control line signals and test line signals. The signal amplitudes of the latter two are much larger than those of the background signals. Therefore,

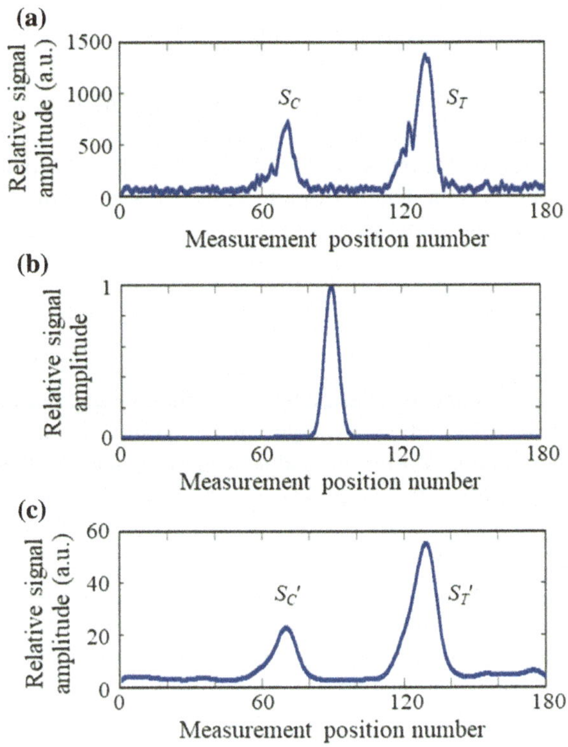

Fig. 5.33 Cross-correlation processing result of input signal with reference signal. Signal after preliminary smoothing filtering (**a**). Gaussian function reference signal (**b**). Cross-correlation processing result (**c**). (**a**) and (**c**) are reproduced from Xie et al. (2009), with permission from the Editorial Department of Acta Photonica Sinica

the adaptive function line boundary orientation algorithm is used in the program; its flow is shown in Fig. 5.34.

In the positioning algorithm, the counter counts by judging the data difference of the two adjacent points when searching from the peak position of a control line to point 1. If the difference is less than the set threshold, the count value is increased by 1; otherwise, the counter value is reset to 0. When the count value reaches the set value, the point position is determined as the left boundary of the control line. The same method is used in searching the right boundary of the control line from the peak position of the control line to that of the test line. The determination process of the test line boundary positions is similar.

The ability of the algorithm to automatically position the boundaries of the function lines is shown in Fig. 5.35. It should be noted that the data used in the adaptive function line boundary orientation algorithm are the data after cross-correlation processing, but the data used to calculate the detection result after removing the noise are the data after preliminary smoothing. Therefore, the signal processing algorithm may not affect the accuracy of the ultimate detection result.

3. Calibration

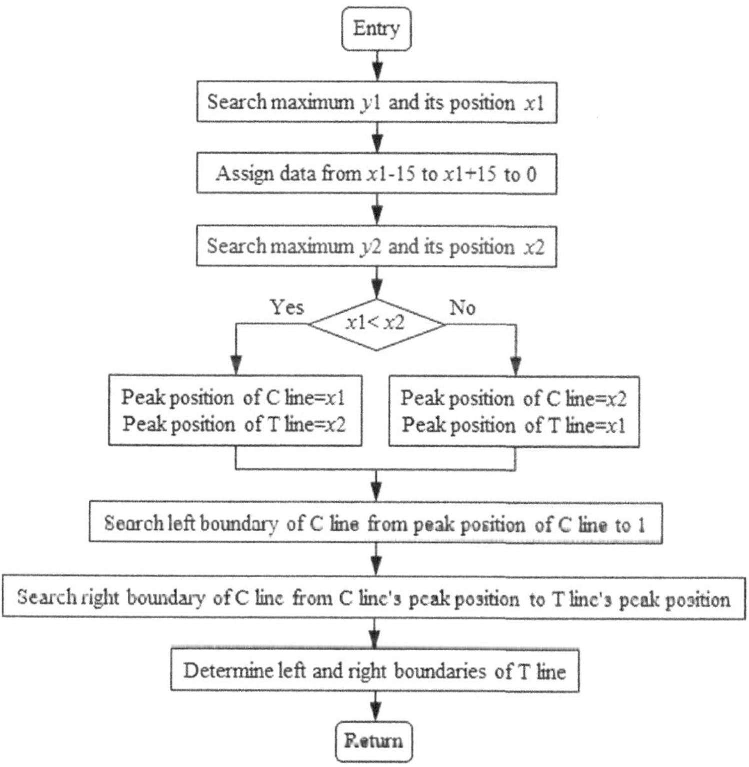

Fig. 5.34 Flow chart of adaptive function line boundary orientation algorithm

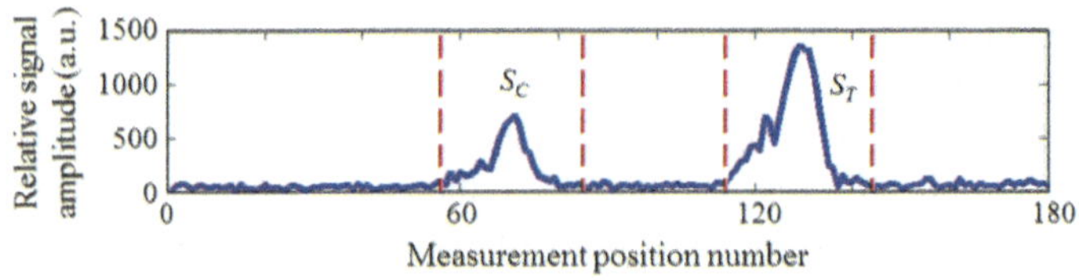

Fig. 5.35 Result of adaptive function line boundary orientation (Xie et al. 2009). Reproduced with permission from the Editorial Department of Acta Photonica Sinica

First, the detection repeatability (Group 1) and the validity of the function line orientation algorithm (Group 2) of the UPT-3A analyzer were evaluated using a double-antigen sandwich LF strip added with a 250 ng/ml plague F1 antibody standard sample (Fig. 5.36). In addition, the LF strip was repeatedly measured with a changing PMT gain control to evaluate the validity of using S_T/S_C as the detection result of the analyzer (Group 3). Each of the above three specifications was evaluated by the *CV* value of 10 measurement results. The *CV* values of the three groups were all less than 5%, which indicates that the UPT-3A analyzer had good detection repeatability and reliable performance, and met the requirements of practical applications.

The detection results of a series of 10 plague F1 antibody standard samples with different concentrations using double-antigen sandwich LF strips and the UPT-3A analyzer are shown in Table 5.7, and the average of three detection results was taken as the ultimate result for each LF strip.

Using the linear fitting to the detection results within the concentration range from 0 to 450 ng/ml, the linear equation obtained is $Y = 0.00801 \times X + 0.88856$, where X is the concentration of the sample, and Y is the S_T/S_C value (Fig. 5.37). r of the fitting equation is 0.97094, indicating that the response linearity of the UPT biosensor in the concentration range was good.

A photograph of the developed UPT-3A analyzer is shown in Fig. 5.38.

Fig. 5.36 Evaluation results of detection reliability of UPT-3A analyzer

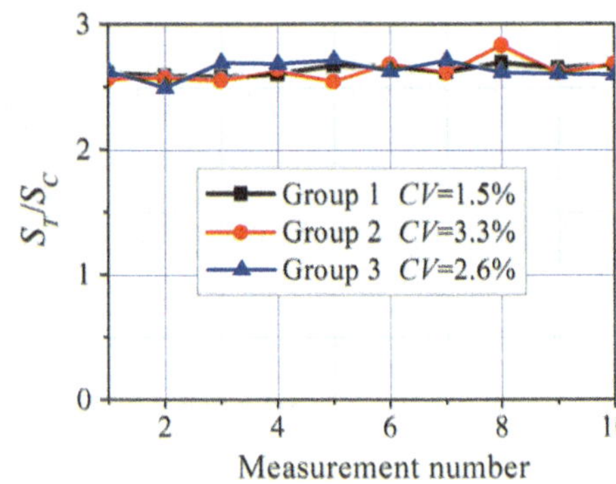

Table 5.7 Detection results of plague F1 antibody standard samples with different concentrations using a UPT-3A analyzer

LF strip number	Sample concentration (ng/ml)	S_T/S_C			
		First	Second	Third	Average value
1	0	0.39	0.39	0.38	0.387
2	50	1.16	1.19	1.17	1.173
3	100	2.01	1.97	1.92	1.967
4	150	2.52	2.59	2.53	2.547
5	200	2.63	2.58	2.46	2.557
6	250	2.79	2.77	2.78	2.780
7	300	3.70	3.66	3.68	3.680
8	350	3.68	3.65	3.55	3.627
9	400	3.93	3.90	3.93	3.920
10	450	4.42	4.23	4.15	4.267

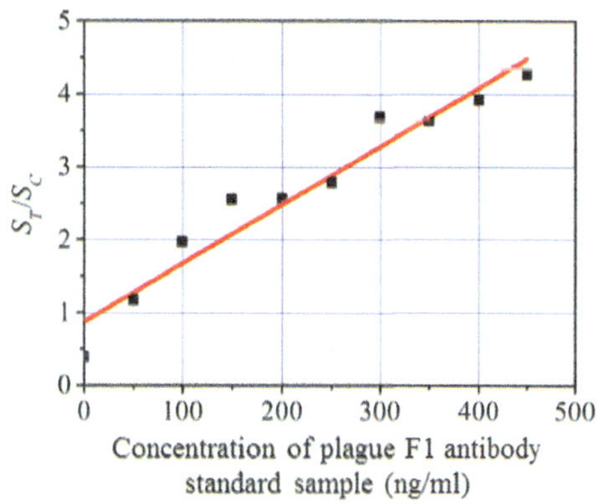

Fig. 5.37 Detection results of plague F1 antibody standard samples with different concentrations using UPT-3A analyzer and fitted response characteristic curve

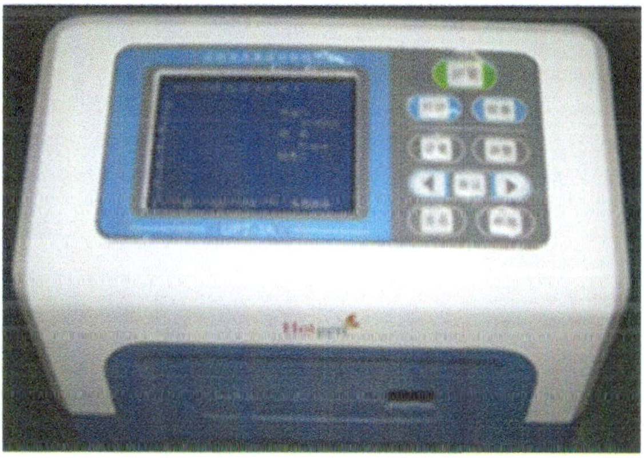

Fig. 5.38 UPT-3A analyzer

5.4.2 Development of Multichannel UPT Immunoassay Analyzer

The UPT-M analyzer is a multichannel UPT immunoassay analyzer that uses a 10-channel UPT lateral flow disc (hereafter called a TC-UPT-LF disc) as the detection object. The analyzer can detect ten target analytes through a single operation of one sample in two minutes, which greatly improves the detection efficiency.

A photograph of the TC-UPT-LF disc loaded with the LF strips is shown in Fig. 5.39. The TC-UPT-LF disc is a regular decagon structure consisting of a base and an upper cover (Hong et al. 2010). Ten LF strips are placed into the corresponding strip-holding channels on the base that are in the directions of the connecting lines between the center point and the vertexes of the regular decagon. The angle interval between the adjacent two LF strips is 36°. Each LF strip is composed of a sample pad, conjugate pad, analytical membrane and absorbent pad wherein a test line and a control line are disposed on the analytical membrane. Drainage channels, result-scanning windows and end-index windows are located at positions corresponding to the LF strips on the upper cover of the TC-UPT-LF disc, and a sample-adding window is located at the center of the upper cover. A drainage piece overlapping the sample pad of each LF strip exists between the LF strips and the upper cover to provide an even flow of added sample to each LF strip.

The UPT-M analyzer consists of a photoelectric measurement system, control system, algorithm module, mechanical structure and other key units. The working principle is shown in Fig. 5.40.

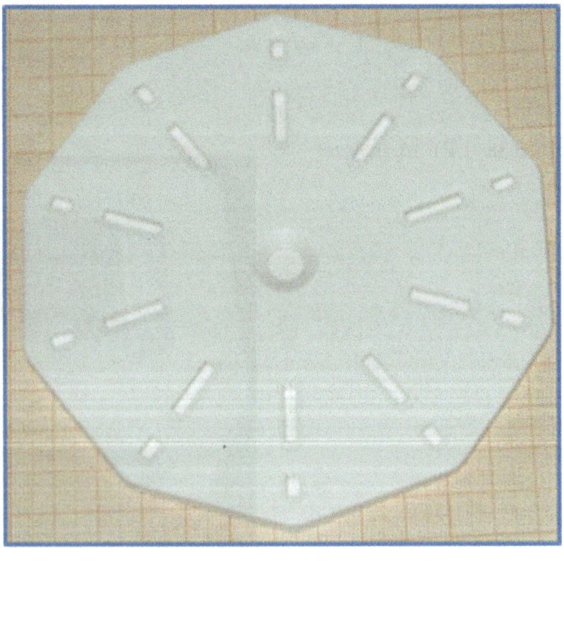

Fig. 5.39 TC-UPT-LF disc loaded with LF strips

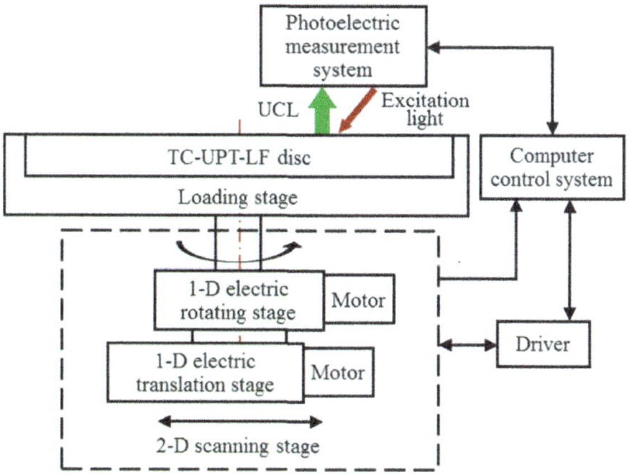

Fig. 5.40 Schematic of working principle of UPT-M analyzer

5.4.2.1 Photoelectric Measurement System

The photoelectric measurement system of the UPT-M analyzer adopts an optical system with an off-axis structure. The axis of the light receiving unit is perpendicular to the TC-UPT-LF disc surface, and has an angle of 45° with the optical axis of the illumination unit.

5.4.2.2 Control System

The control system hardware of the UPT-M analyzer mainly consists of a PC/104 embedded SBC and a multifunctional data acquisition card. Its main functions include the control of the translational motion and the rotational motion of the two-dimensional scanning stage, the acquisition of the voltage signal output from the photoelectric measurement system and the output of the detection result.

The program flow of the control system is shown in Fig. 5.41. A multi-thread technique is used, three auxiliary threads are opened outside the main thread. The main thread captures a user's real-time input message, and draws the data curve. Thread 1 outputs the pulse signal and the direction signal to the motor drivers, acquires the voltage signal, and outputs the data to the main thread. Thread 2 reads the RFID data, and thread 3 prints the detection results.

5.4.2.3 Algorithm Module

The algorithm module includes the motion control algorithm, signal acquisition algorithm and data processing algorithm. The motion control algorithm implements the translational and rotational motions of the two-dimensional scanning stage, and the difficulty is to meet the positioning accuracy requirements. The data processing algo-

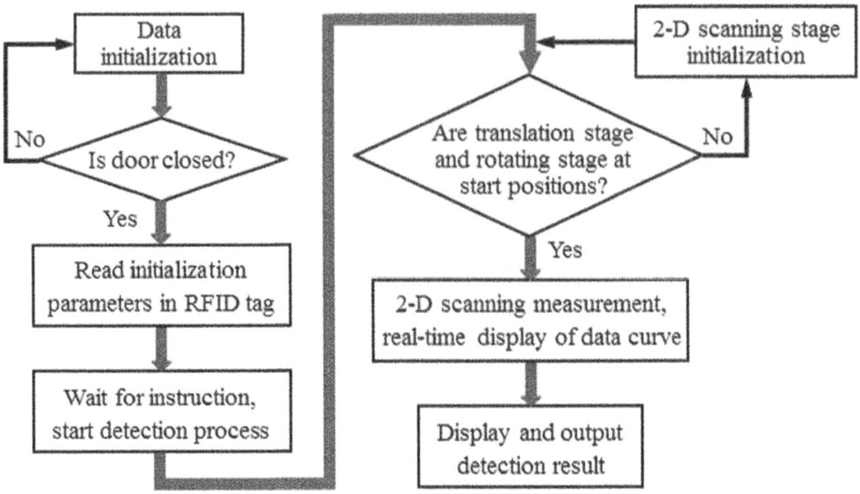

Fig. 5.41 Program flow chart of UPT-M analyzer control system

rithm includes the smoothing and denoising of the data curve of a single LF strip, automatic position searches and signal peak area calculations of the test line and the control line. The specific content is similar to that of the UPT-2 analyzer.

5.4.2.4 Calibrartion

Three TC-UPT-LF discs were measured 10 times using the UPT-M analyzer, and the *CV* values of the detection results of most LF strips were better than 2%, those of the other individual LF strips were between 3% and 5%. These results indicate that the analyzer had good detection repeatability and reliable performance.

A photograph of the developed UPT-M analyzer and the user interface are shown in Figs. 5.42 and 5.43, respectively.

5.4.3 Development of Imaging Detection System for LF Strip

The imaging detection system for the LF strip (hereafter called the imaging detection system) mainly includes a photoelectric measurement system, a control system consisting of a control circuit and a signal acquisition and processing circuit (Fig. 5.44).

The photoelectric measurement system is composed of a light source, an optical system consisting of an illumination unit and an an imaging unit and a linear image sensor. The optical system adopts an off-axis structure. The collimated laser beam emitted from an LD module with a 980-nm wavelength, i.e., the light source, forms a uniform light spot which covers the test and control lines on an LF strip through

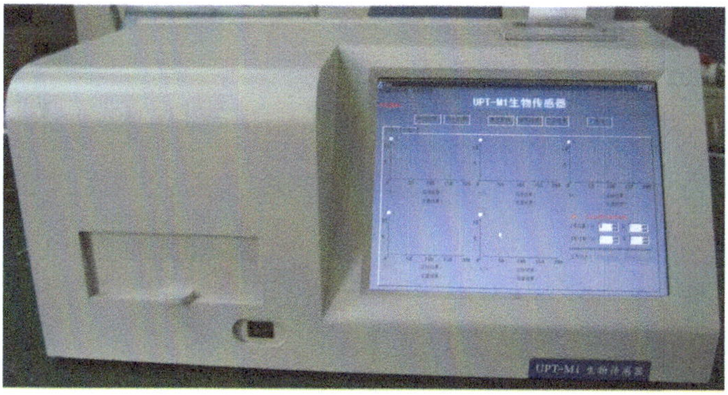

Fig. 5.42 UPT-M analyzer

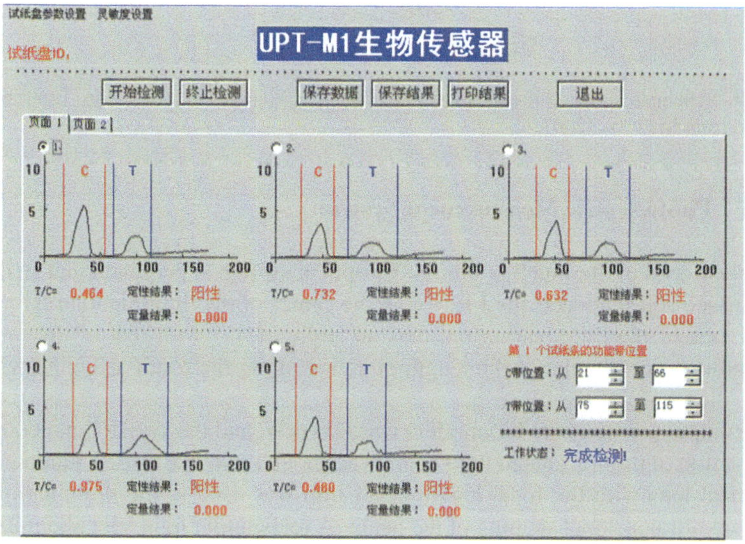

Fig. 5.43 UPT-M1 analyzer user interface

the illumination unit. The distribution of the UCL signal within the measurement bandwidth generated by the UCP particles in the illumination region is directly imaged on the linear image sensor through the imaging unit. The control system drives the linear image sensor, acquires its output signal and performs processing and analysis, and obtains the detection result (Liu et al. 2007).

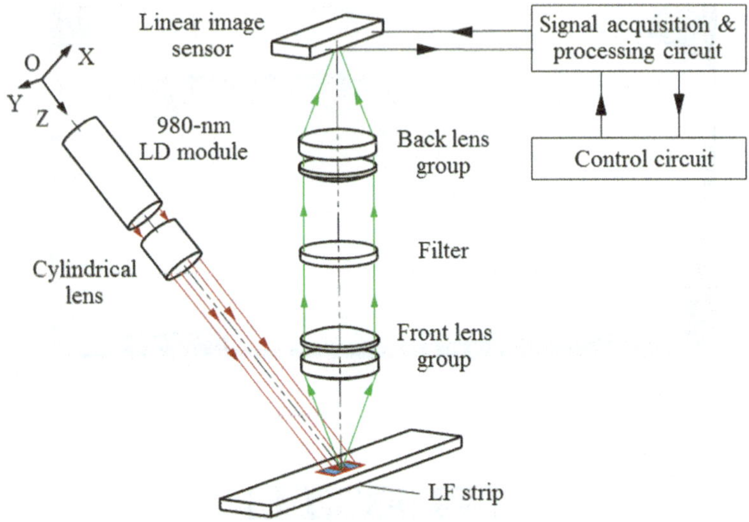

Fig. 5.44 Schematic of working principle of imaging detection system for LF strip. Translated and modified from Liu et al. (2007)

5.4.3.1 Photoelectric Measurement System

The optical axis of the imaging unit in the optical system is perpendicular to the LF strip surface and intersects the LF strip at the center of the illumination region. The optical axis of the illumination unit has an angle of 45° with that of the imaging unit, and is perpendicular to the length direction along the result-scanning window (Y-axis direction) of the LF strip.

The output power of the LD module was 800 mW, and the size of the rectangular cross section of the collimated light beam was about 4 mm × 6 mm. A plano-convex cylindrical lens with the focal length of 71 mm was used in the illumination unit for the one-dimensional shaping of the beam. A rectangular light spot about 2.5 mm × 6 mm was formed on the LF strip surface by adjusting the distance between the cylindrical lens and the LF strip.

The imaging unit adopting a symmetrical structure consisted of a front lens group, a filter and a back lens group. The imaging magnification of the imaging unit was −1×, and the field of view was ϕ6.4 mm.

An S3924-256Q NMOS linear image sensor (Hamamatsu Photonics KK) was used behind the imaging unit to receive the one-dimensional distribution image of the UCL. The photosensitive surface size of the sensor was 2.5 mm × 6.4 mm. The sensor had a total of 256 pixels, and the width of a single pixel was 25 μm. The total photosensitive surface could receive the UCL signals of the entire illumination region on the LF strip, and each pixel corresponds to an area with a width of 25 μm (Hamamatsu Photonics KK 2014).

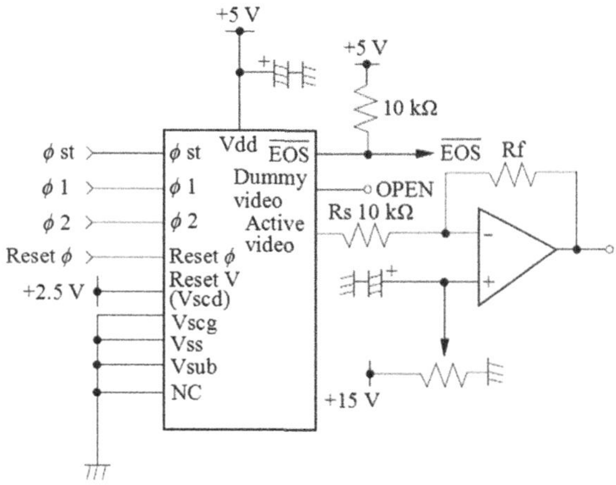

Fig. 5.45 Schematic of signal readout circuit of S3924-256Q NMOS linear image sensor (Hamamatsu Photonics KK 2014)

In addition, the voltage signals of the S3924-256Q NMOS linear image sensor can be read out by a simple external circuit (Fig. 5.45).

5.4.3.2 Control System

The control system hardware of the imaging detection system mainly consists of two parts: the control circuit and the signal acquisition and processing circuit. It adopts a dual SCM mode, i.e., a master SCM in the control circuit and a slave SCM in the signal acquisition and processing circuit.

The signal acquisition and processing circuit mainly drives the NMOS linear image sensor, acquires the output signal, and stores and processes data. The circuit is composed of the slave SCM, external RAM, integrated operational amplifier, A/D converter, NMOS linear image sensor, address flip-latch, external I/O port and RS232 interface (Fig. 5.46).

The measurement process is as followings: (1) The drive pulses are output from the I/O port of the slave SCM to the image sensor; (2) the output signal from the image sensor is amplified by the integrated operational amplifier, and quickly acquired and

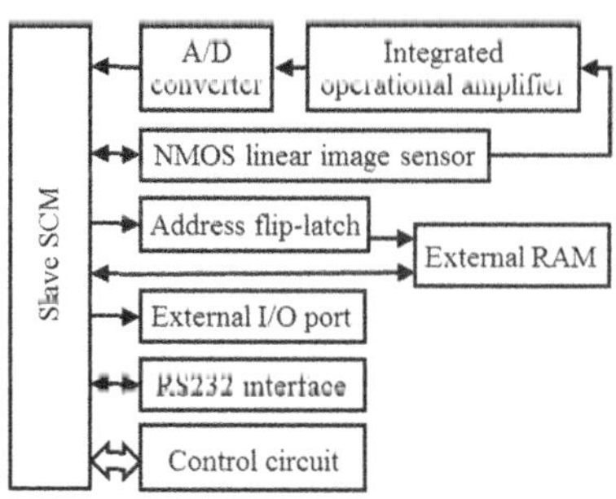

Fig. 5.46 Block diagram of signal acquisition and processing circuit. Translated and modified from Liu et al. (2007)

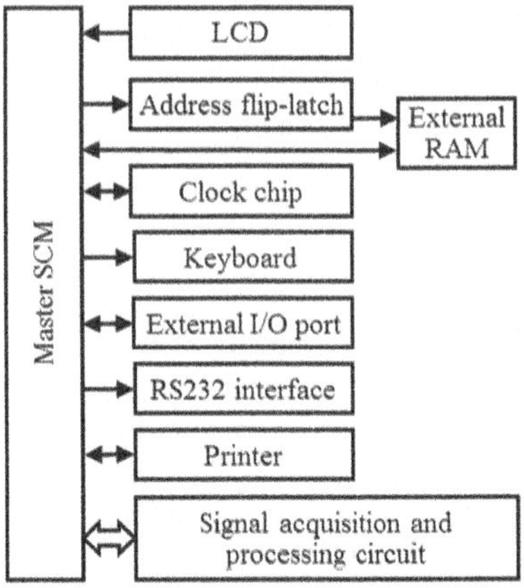

Fig. 5.47 Block diagram of control circuit. Translated and modified from Liu et al. (2007)

converted by the A/D converter; (3) the DC component contained in the output signal is filtered out by an adjustable DC voltage at the noninverting input of the differential amplifier; (4) when the conversion of all the signals is completed, the slave SCM processes and analyzes the obtained data, calculates the positions of the control line and test line, sends the corresponding signal amplitudes to the external I/O port, and waits for the control circuit to receive and process the data.

The control circuit is mainly composed of the master SCM, clock chip, external RAM, external I/O port and RS232 interface (Fig. 5.47). The control circuit controls the operation of the detection system and displays the detection results. Dialog boxes displayed on the LCD monitor prompt users to input parameters and select instructions with the keyboard. When the user selects the signal acquisition command, the master SCM sends a signal to the slave SCM in the signal acquisition and processing circuit, waits for acquisition-complete feedback from the slave SCM, and receives the results data from the slave SCM through the external I/O port. After the data is processed by the master SCM, the detection result, detection time, target analyte ID number and other information is displayed on the monitor and can be printed.

5.4.3.3 Algorithm Module

The flow of the data processing algorithm of the imaging detection system is shown in Fig. 5.48, and the content is similar to that of the scanning analyzer, including the smoothing filter, adaptive function line boundary positioning, linear denoising, summing calculation and so on.

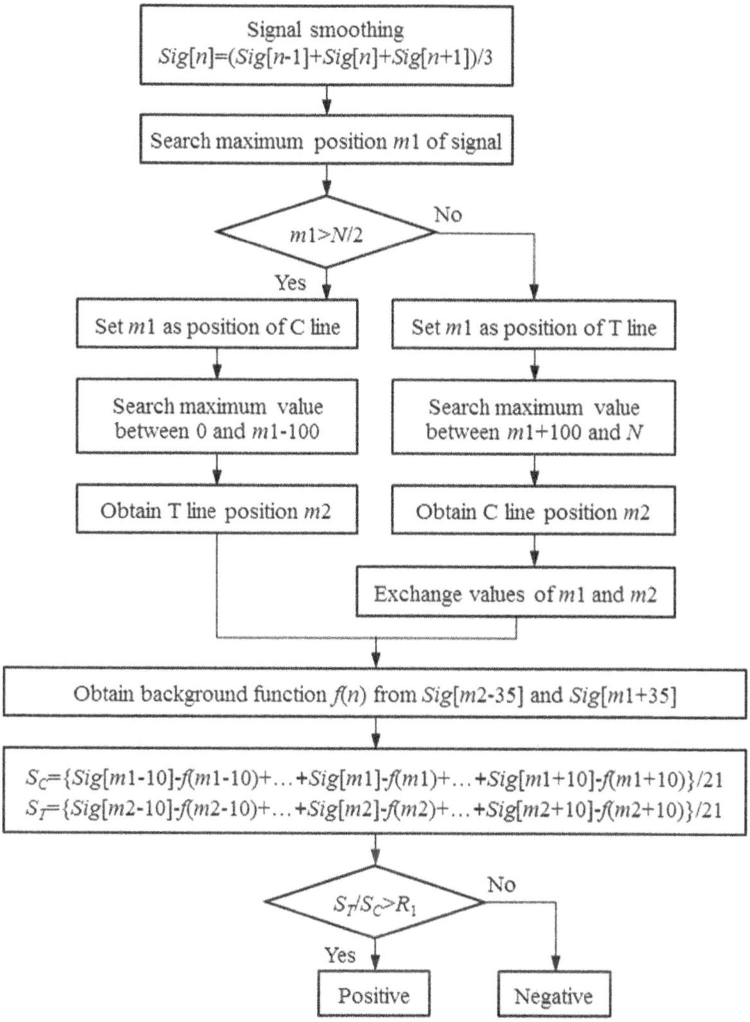

Fig. 5.48 Flow chart of data processing algorithm. Translated and modified from Liu et al. (2007)

5.4.3.4 System Calibration

The UCL signal distribution curve of a double-antigen sandwich LF strip added with 10 ng/ml rabbit anti-plague IgG standard sample was measured by the imaging detection system (Fig. 5.49). It can be seen that the signals within the test line and the control line in the curve are prominent, but the background noise is relatively high.

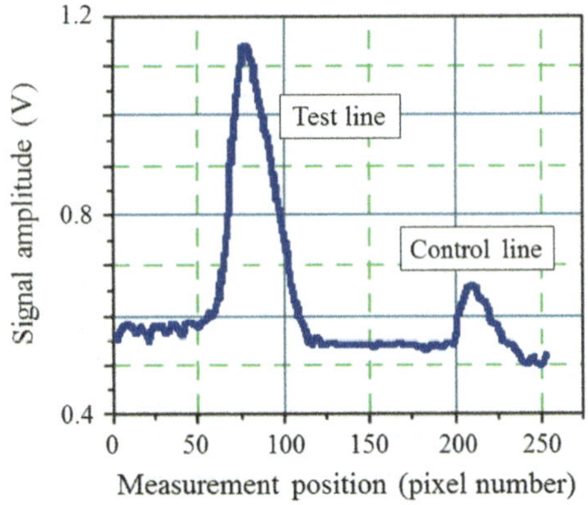

Fig. 5.49 UCL signal distribution measured by imaging detection system on LF strip

The detection repeatability of the imaging detection system was evaluated using a double-antigen sandwich LF strip of another batch which was added with 10-ng/ml rabbit anti-plague IgG standard sample (Fig. 5.50). The detection system performed 10 measurements.

The *CV* value of the S_T/S_C in the evaluation results was 8.81%, less than 10%, indicating that the detection repeatability of the imaging detection system was relatively good but not ideal. The reasons can be explained as follows:

(1) The illumination region size of the detection system was relatively large and the light intensity distribution was uneven. To reach the required power density, the power of the laser needed was high, and therefore the heat was large, the performance was unstable.
(2) The spectral responsivity within the measured UCL bandwidth of the NMOS linear image sensor used in the detection system was lower than that of the photoelectric converter used in the scanning analyzer, while the spectral responsivity within the wavelength range of the illumination light was high. Therefore, the detection system was easily disturbed by stray light, and had a low S/N ratio.

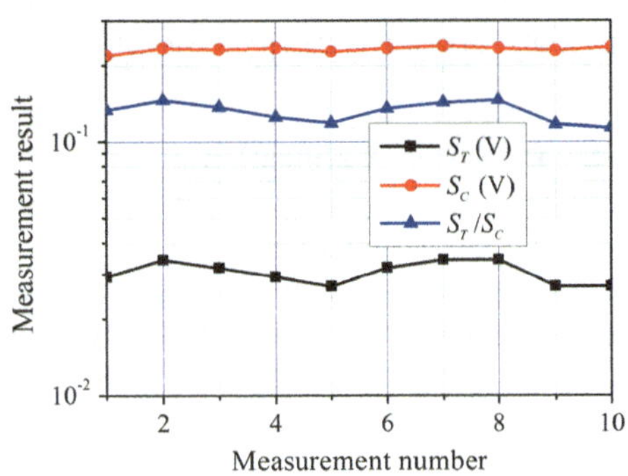

Fig. 5.50 Evaluation results of detection repeatability of imaging detection system

Table 5.8 Detection results of rabbit antiplague IgG standard samples with different concentrations using imaging detection system

LF strip number	Sample concentration (ng/ml)	S_T (V)	S_C (V)	S_T/S_C
1	0	0.05302	0.24074	0.22024
2	2	0.02363	0.09508	0.24857
3	5	0.1024	0.16558	0.31846
4	10	0.06066	0.18055	0.33595
5	20	0.09027	0.1852	0.48745
6	30	0.08547	0.09989	0.85568
7	50	0.08287	0.04679	1.77098
8	60	0.13407	0.05813	2.3063

Translated and modified from Liu et al. (2007)

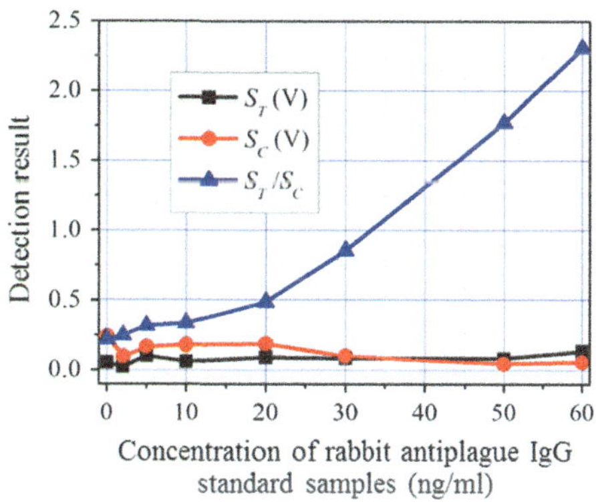

Fig. 5.51 Detection results of rabbit anti-plague IgG standard samples with different concentrations using imaging detection system. Revised from Liu et al. (2007)

The detection results of a series of rabbit anti-plague IgG standard samples with different concentrations using the double-antigen sandwich LF strips and the imaging detection system are shown in Table 5.8 and Fig. 5.51.

The linear fitting is performed to the detection results within the concentration range. The correlation coefficient r is 0.97599. It can be seen that the imaging detection system had a good response linearity within the concentration range of 0–60 ng/ml and was able to perform quantitative detection.

5.5 Conclusions

The UPT biosensor is an immunochromatographic optical biosensor using UCP particles as the label, and usually consists of a UPT LF strip and a UPT immunoassay analyzer. By measuring the UCL signal distribution of UCP particles on the UPT LF strip, the analyzer obtains the target analyte concentration of the sample on the basis of the response characteristic curve.

In this chapter, the basic concepts of the UPT biosensor were introduced. The working principle of the UPT immunoassay analyzer and the design methods of the four key units, including the photoelectric measurement system, control system, algorithm module and mechanical structure, and the calibration technique were discussed. Finally, the analyzer engineering was presented in detail.

The detection limit of the series of UPT biosensors that we developed for the plague F1 antigen reaches the level of ng/ml and even sub ng/ml. For the UPT-3A analyzer, the correlation coefficient *r* of the linear fitting equation in the concentration range from 0 to 450 ng/ml reached 0.97094.

We completed the development of the principle prototype, the small volume production instrument and the product. These instruments have been successfully applied in the on-site detection of plague in plague foci, rapid on-site detection of anti-bioterrorism at the frontier port, major activity security and clinical diagnosis.

The achievements of research, development and application of the UPT biosensing technology have promoted the development and the integral level improvement of the point-of-care testing (POCT) technology, pioneered the independent research and development of the POCT detection technology and products, and provided a demonstration effect on the development of the innovative POCT detection technology in China.

Bibliography

Byfield MP, Abuknesha RA. Biochemical aspects of biosensors. Biosens Bioelectron. 1994;9:373–400.

Hamamatsu Photonics KK. Photomultiplier tube modules. 2008.

Hamamatsu Photonics KK. NMOS linear image sensor S3921/S3924 series. 2014.

Hampl J, Hall M, Mufti NA, et al. Upconverting phosphor reporters in immunochromatographic assays. Anal Biochem. 2001;288(2):176–87.

Hong W, Huang L, Wang H, et al. Development of an up-converting phosphor technology-based 10-channel lateral flow assay for profiling antibodies against Yersinia pestis. J Microbiol Methods. 2010;83(2):133–40.

Huang L, Zhou L, Zhang Y, et al. A simple optical reader for upconverting phosphor particles captured on lateral flow strip. IEEE Sens J. 2009;9(10):1185–91.

Huang H, Zhao Y, Huang L, et al. Optical biosensors. OE Prod News. 2010;1(8):21–3 (in Chinese).

Ji L, Zhang H. Research and implementation of signal detection based on cross correlation. J Jilin Teach Inst Eng Technol (Eng Technol Ed). 2004;20(6):39–41 (in Chinese).

Li Y, Yang R. Rapid test of immunogold with membrane as solid phase carrier. Prog Microbiol Immunol. 2003;31(1):74–8 (in Chinese).

Liu L, Zhou L, Huang L, et al. CCD-based detection system for immunity-chromatography test strip. Chin J Sci Instrum. 2007;28(2):246–51 (in Chinese).

Lu J, Zhou L, Zhao Y, et al. Scanning detection system of UPT-based immunoassay test strip. Acta Photonica Sin. 2006;35(4):555–60 (in Chinese).

Menyuk N, Pierce JW, Dwight K. $NaYF_4$: Yb, Er—efficient upconversion phosphor. Appl Phys Lett. 1972;21(4):159–61.

Miao Y. Biosensors and their military applications. Beijing: National Defense Industry Press; 2005 (in Chinese).

Niedbala RS, Vail TL, Feindt H, et al. Multiphoton up-converting phosphors for use in rapid immunoassays. Proc SPIE. 2000;3913:193–203.

Niedbala RS, Feindt H, Kardos K, et al. Detection of analytes by immunoassay using up-converting phosphor technology. Anal Biochem. 2001;293(1):22–30.

Page RH, Schaffers KI, Waide PA, et al. Upconversion-pumped luminescence efficiency of rare-earth-doped hosts sensitized with trivalent ytterbium. J Opt Soc Am B. 1998;15(3):996–1008.

Si S. Biosensors. Beijing: Chemical Industry Press; 2003 (in Chinese).

Suyver JF, Aebischer A, Biner D, et al. Novel materials doped with trivalent lanthanides and transition metal ions showing near-infrared to visible photon upconversion. Opt Mater. 2005a;27(6):1111–30.

Suyver JF, Aebischer A, García-Revilla S, et al. Anomalous power dependence of sensitized upconversion luminescence. Phys Rev B. 2005b;71(12):125123.

Suyver JF, Grimm J, Krämer KW, et al. Highly efficient near-infrared to visible up-conversion process in $NaYF_4$: Er^{3+}, Yb^{3+}. J Lumin. 2005c;114(1):53–9.

Suyver JF, Grimm J, van Veen MK, et al. Upconversion spectroscopy and properties of $NaYF_4$ doped with Er^{3+}, Tm^{3+} and/or Yb^{3+}. J Lumin. 2006;117(1):1–12.

Xie C, Zhang Y, Huang L, et al. Signal processing algorithm on up-converting phosphor technology-based biosensor. Acta Photonica Sin. 2009;38(12):3256–60 (in Chinese).

Yan Z, Zhou L, Zhao Y, et al. Rapid quantitative detection of Yersinia pestis by lateral-flow immunoassay and up-converting phosphor technology-based biosensor. Sens Actuators B. 2006;119:656–63.

Zhang X. Biosensors. Beijing: Chemical Industry Press; 2006 (in Chinese).

Zhao S, Hou Y, Sun L. Studies on the upconversion luminescence of a new material doped with Er^{3+}. J North Jiaotong Univ. 1999;23(6):46–9 (in Chinese).

Zhao S, Hou Y, Dong J. Upconversion luminescence of rare earth ions. Semicond Optoelectron. 2000;21(4):241–4 (in Chinese).

Zhao S, Xu Z, Pei X. Donor concentration dependence of Upconversion luminescence in $YLiF_4$: Er^{3+}, Yb^{3+}. Spectrosc Spectr Anal. 2005a;25(12):1933–7 (in Chinese).

Zhao Y, Zhou L, Huang H, et al. Up-converting phosphor technology-based biosensor and its application. Acta Optica Sin. 2005b;25(6):841–7 (in Chinese).

Zhao Y, Zhou L, Wang J, et al. Research on optical biosensor with up-converting phosphor marker. Chin Opt Lett. 2006;4(8):464–6.

Zhou L, Ji J, Yang R. Up-converting phosphor technology in the rapid bioanalysis. Biotechnol Bull. 2003;3:20–5 (in Chinese).

Chapter 6
Industrialization of Up-Converting Phosphor Diagnostic Products

Changqing Lin

Abstract The key steps in UPT technology industrialization include the combination of UPT technology and immunochromatography, the efficient binding of UCNPs and biomolecules, the uniformity of coated PVC sheets, the scale-up of UCNPs particle preparation, and large-scale production of UPT biosensors. In the field of UPT diagnostic production, 26 medical device registration certificates have been approved by the China Food and Drug Administration. The products are widely applied in disease prevention, clinical emergency treatment and first aid, and military and public security.

Keywords Industrialization · Medical device registration certificates · Large-scale production · Market demand · Point-of-care testing

Industrialization of immunochromatographic diagnostic reagents based on up-converting phosphor technology (UPT) depends on a number of factors, including; the large-scale preparation and chemical modification of rare earth nanoparticles with up-conversion luminescence properties (up-converting phosphor nanoparticles, UCNPs); the efficiency of the combination of the UCNPs and chosen biomaterials (antibodies and antigens); overcoming the agglomeration of nanomaterials in large-scale production; the large-scale stability and reliable production of test strips; and the large-scale production of UPT biosensors. China's extensive market for diagnostic reagents and the growing healthcare needs of the general population, are important supports for the industrialization of this technology.

C. Lin (✉)
Beijing Key Laboratory of POCT for Bioemergency and Clinic (No. BZ0329), Beijing 100071, China
e-mail: hotgen@hotgen.com.cn

Beijing Hotgen Biotechnology Co., Ltd., Beijing Z-Park Daxing Biomedical Industry Base, No. 9, Tianfujie, Beijing 102600, China

R. Yang (ed.), *Principles and Applications of Up-converting Phosphor Technology*,
https://doi.org/10.1007/978-981-32-9279-6_6

6.1 Market Demand for Point-of-Care Testing Dictates the Industrialization of UPT Immunochromatographic Diagnostic Reagents

6.1.1 Status of Point-of-Care Testing

Point-of-care testing (POCT) is the rapid detection of unknown organisms, biomarkers or environmental contaminants using portable instruments or reagents, providing basic data to hazard prevention departments and clinicians. POCT is primarily used in high pathogenicity microorganism emergencies, food poisoning emergencies, clinical emergency treatment and first aid, and diagnosis at the bedside. It provides technological support for disease prevention and control departments; military and public security departments; food safety and drug abuse detection departments; hospitals; and quarantine departments for epidemic prevention after natural disasters such as earthquakes, floods, and cement streams. POCT is a high-tech integration of optical electromechanical instruments, new materials, biotechnology, and information science.

Currently, the leading POCT products are characterized by quantitative detection. The fourth generation of POCT products will be characterized by automation and informationization (Melo et al. 2011). In 1995, the National Committee for Clinical Laboratory Standards (NCCLS) of the USA published the AST2-P document, marking the beginning of the standardized management of in vitro diagnostic (IVD) tests. POCT is the fastest-growing segment of the global IVD market, with a market capitalization of billions of dollars (Melo et al. 2011). The automatic POCT instrument for detection of cardiac markers made by the Mitsubishi Corporation (Japan), a fourth-generation POCT instrument, has been on the market since 2011 and has taken the lead in the Chinese market. The majority of the manufacturing sites for low-end products based on colloidal gold immunochromatography—mainly early pregnancy test strips—are in China and India. The added value of the products is very low, and the number of production enterprises is in the hundreds. In China, the production value of colloidal gold immunochromatography reagents is only ~1 billion yuan, and the market is no longer growing. In 2011, the National High Technology Research and Development Program of China began to support enterprises for research into POCT clinical application and industrialization. As part of the "11th Five-Year Plan of China," The National Science and Technology Major Project began providing funds for the study of infectious disease diagnosis; however, it was not until 2012 that emergency management and POCT were officially listed in the "12th Five-Year Plan" for the medical device technology industry. UPT immunochromatographic diagnostic systems have been developed through the collaboration of Beijing Institute of Microbiology and Epidemiology (Beijing, China), Shanghai Kerune Phosphor Technology Co., Ltd. (Shanghai, China), Shanghai Institute of Optics and Fine Mechanics of the Chinese Academy of Sciences (Shanghai, China), and Beijing Hotgen Biotechnology Co., Ltd. (Beijing, China). The current UPT diagnostic system is the third-generation POCT products, subsequent development of handheld intelligent UPT instruments will allow China to fall into line with international advancements.

6.1.2 *Market Demand for POCT*

6.1.2.1 Application of POCT in Disease Prevention

The POCT detection method is critical for the control of various infectious diseases, such as human immunodeficiency virus (HIV) (Fonjungo et al. 2016), Ebola virus (Kost 2018), Plasmodium (Keitel et al. 2017), and syphilis (Young et al. 2018). After the Haiti earthquake in 2010, cholera was epidemic for a long period of time owing to outdated detection technology and ineffective anti-epidemic measures; resulting in the deaths of thousands of people. During the Olympic Games in 2008, the Ministry of Health of China issued a report detailing an emergency laboratory detection plan for highly pathogenic microorganisms for the 100 days of the Olympic Games, and prepared a variety of emergency response plans related to highly pathogenic microorganisms (such as *Y. pestis*, *B. anthracis*, *Brucella* spp., *Salmonella typhi*, and *Vibrio Cholerae*), effectively guaranteeing the biosecurity of China during the Olympic Games.

6.1.2.2 Application of POCT in Clinical Emergency Treatment and First Aid

Clinical POCT is leading a major revolution in laboratory medicine. It is rapid, convenient, efficient, low-cost, uses small samples, and does not require professional operation. The reliable results can be directly and quickly obtained without complex specimen pretreatment, numerous pieces of cumbersome testing equipment, or data processing. In addition, its reagents are stable, easy to store and transport, and can be widely used in emergency rooms, clinical laboratories, community primary hospitals, and in the home. A series of POCT products for heart disease (such as myocardial infarction and acute heart failure) have been widely used in clinical practice. The 2010 bulletin of the leading POCT company, BIOSITE (USA), pointed out that their products are primarily focused toward the clinical diagnosis of cardiovascular and cerebrovascular diseases and drug abuse. New POCT products for the prevention of poor health in the community make up an emerging field with a broad potential market (He et al. 2018), while precision medicine will also promote the rapid development of POCT in the future (Ford Carleton et al. 2016).

6.1.2.3 Application of POCT in the Military and Public Security

The anthrax spore attack that followed the September 11th (9.11) terrorist incidents in the United States, generated a high degree of panic, causing the US government to make biological terrorism incidents a high priority and to formulate three anti-bioterrorism plans. After the 9.11 incidents, the Chinese Ministry of Science and Technology also launched a special project for rapid pathogen detection and identification (50 million yuan). Since 2006, the Ministry of Science and Technology and the Ministry of Public Security have jointly launched a new anti-bioterrorism research project. From 2009, the major Chinese National Science and Technology Programs for infectious disease also supported research on biological rapid detection and identification. In 2011, the implementation program for the fire department introduced devices related to biological emergency testing to enhance their capabilities.

6.2 The Key Steps in UPT Technology Industrialization

6.2.1 Step One: The Combination of UPT Technology and Immunochromatography

UPT technology combines up-converting phosphor nanoparticles (UCNPs) with a variety of bioactive molecules, and then uses miniaturized optoelectronic materials to develop sensitive, fast, and easy-to-use biosensors, which can be applied for biomedical detection. Currently, the UPT instrument developed by the US Department of Defense is used for biological anti-terrorism to quantitatively detect *Y. pestis* and *B. anthracis* spores, while that developed by OraSure Technologies (USA) is used to detect antibodies against HIV and for quantitative detection of drug abuse. Beijing Institute of Microbiology and Epidemiology, China's domestic research institute, has been active in this field since 2000. They have successfully developed UPT technology-based immunochromatographic reagents for detection of pathogens such as *Y. pestis*, *B. anthracis* spores, and *E. coli* O157, among others. More than 40 patents have been authorized and 20 related papers have been published, while test kits have been produced by Beijing Hotgen Biotechnology Co., Ltd. for industrialization.

6.2.2 Step Two: The Efficient Binding of UCNPs and Biomolecules

The coupling of nanoparticles and biomolecules is conventionally through physical adsorption (colloidal gold) and chemical coupling (latex or fluorescent particles); however, UCNPs of rare earth materials are difficult to effectively label with biomolecules using conventional coupling methods. Experiments have shown that the efficiency of combining UCNPs with biomolecules has been improved by 100–700% $\pm$ 30%, while the detection sensitivity now reaches the pg level with excellent specificity. The yield of a single batch can be used for the fabrication of more than 24,000 strips, realizing high-efficiency coupling of UCNPs particles and biomolecules and large-scale preparation of UCNPs-biomolecule complexes.

6.2.3 Step Three: The Uniformity of Coated PVC Sheets

After the PVC support is coated, it is dried in a uniform environment to ensure a consistent performance between the test strips. The amount of PVC sheet coated is limited by the use of a conventional drying oven. A drying system was therefore established to expand the scale of a single batch to more than 70,000 strips, demonstrating successful scale-up of coated PVC support production.

6.2.4 Step Four: The Scale-Up of UCNPs Particle Preparation

The amount of particles in one batch prepared in a conventional laboratory can be in the hundreds of milligrams range, but rarely exceeds 1 g. Through process optimization, the UCNPs particle output of a single batch can be increased to several tens of grams, which provides a solid foundation for the industrialization of UPT kits.

6.2.5 Step Five: Large-Scale Production of UPT Biosensors

The large-scale production of UPT biosensors plays an important role in guaranteeing quantitative detection. The UPT biosensor production line was developed by optimizing the analysis software, the optical path of the sensor, and the noise reduction in the quantitative analysis.

6.3 Social and Economic Benefits of Industrialization of UPT Diagnostic Production

6.3.1 Products that Have Obtained the Medical Device Registration Certificate and CE Certification from the China Food and Drug Administration

In the field of UPT diagnostic production, 26 medical device registration certificates—including 23 medical device registration certificates for UPT detection kits and three for UPT biosensors—have been approved by the China Food and Drug Administration (Table 6.1).

Table 6.1 Registration certificates for UPT-related medical devices approved by the China Food and Drug Administration

No.	Production
1	UPT detection kit for C reaction protein (CRP)
2	UPT detection kit for collagen type IV (CIV)
3	UPT detection kit for laminin (LN)
4	UPT detection kit for hyaluronic acid (HA)
5	UPT detection kit for amino terminal peptide of procollagen type III (PIIINP)
6	UPT detection kit for serum tissue inhibitor of metalloproteinase-1 (TIMP-1)
7	UPT detection kit for myocardium troponin I (cTnI)
8	UPT detection kit for N-terminal fragment of B-type natriuretic peptide precursor (NT-proBNP)
9	UPT detection kit for fetal fibronectin (fFN)
10	UPT detection kit for alpha-fetoprotein (AFP)
11	UPT detection kit for golgin 73 (GP73)
12	UPT detection kit for procalcitonin
13	UPT detection kit for heart-type fatty acid-binding protein
14	UPT detection kit for D-dimer
15	UPT detection kit for human plasma lipoprotein-associated phospholipase A2 (Lp-PLA2)
16	UPT detection kit for neutrophil gelatinase-associated lipocalin
17	UPT detection kit for interleukin 6
18	UPT detection kit for antibodies against cyclic citrullinated peptide
19	UPT detection kit for myoglobin (MYO)
20	UPT detection kit for creatine kinase isoenzyme (CK-MB)
21	UPT detection kit for ketamine
22	UPT detection kit for morphine
23	UPT detection kit for methamphetamine
24	UPT biosensor

6.3.2 Social Benefits of UPT Diagnostic Products

Since the successful industrialization of UPT diagnostic reagents in 2011 (Yang et al. 2017), they have been successfully applied in more than 2000 hospitals in China, including Grade-A tertiary hospitals, regional second-class hospitals, community health service stations, and township health centers, among others. The products are widely used in national, provincial, and municipal Centers for Disease Control and Prevention in China. The detection reagents for biological warfare agents and infectious pathogens have been deployed at more than 70 entry–exit inspection and quarantine ports in China. Detection reagents for mycotoxins have also been widely used in large-scale fodder enterprises in China.

Currently, the cumulative production value of UPT diagnostic reagents is 330 million yuan, with a total tax payment of more than 23 million yuan, and the industry has 200 employees. As an advanced manufacturer of detection reagents for medical and public safety, Beijing Hotgen Biotechnology Co., Ltd. has made significant contributions to the industrialization of UPT diagnostic production.

References

Fonjungo P, Boeras D, Zeh C, Alexander H, Parekh B, Nkengasong J. Access and quality of HIV-related point-of-care diagnostic testing in global health programs. Clin Infect Dis. 2016;62(3):369–74.

Ford Carleton P, Schachter S, Parrish J, Collins J, Crocker J, Dixon R, et al. National institute of biomedical imaging and bioengineering point-of-care technology research network: advancing precision medicine. IEEE J Transl Eng Health Med. 2016;4:2800614.

He W, You M, Wan W, Xu F, Li F, Li A. Point-of-care periodontitis testing: biomarkers, current technologies, and perspectives. Trends Biotechnol. 2018.

Keitel K, Kagoro F, Samaka J, Masimba J, Said Z, Temba H, et al. A novel electronic algorithm using host biomarker point-of-care tests for the management of febrile illnesses in Tanzanian children (e-POCT): a randomized, controlled non-inferiority trial. PLoS Med. 2017;14(10):e1002411.

Kost G. Molecular and point-of-care diagnostics for Ebola and new threats: national POCT policy and guidelines will stop epidemics. Expert Rev Mol Diagn. 2018;18(7):657–73.

Melo M, Clark S, Barrio D. Miniaturization and globalization of clinical laboratory activities. Clin Chem Lab Med. 2011;49(4):581–6.

Yang X, Liu L, Hao Q, Zou D, Zhang X, Zhang L, et al. Development and evaluation of up-converting phosphor technology-based lateral flow assay for quantitative detection of NT-proBNP in blood. PLoS ONE. 2017;12(2):e0171376.

Young N, Taegtmeyer M, Aol G, Bigogo G, Phillips-Howard P, Hill J, et al. Integrated point-of-care testing (POCT) of HIV, syphilis, malaria and anaemia in antenatal clinics in western Kenya: a longitudinal implementation study. PLoS ONE. 2018;13(7):e0198784.

Chapter 7
Application of UPT-POCT in Emergency Medicine

Yanzhao Li, Ruifeng Xiao and Yong Zhao

Abstract UPT-POCT technology is a representative technology platform for precision quantitative POCT detection and diagnosis. It has multiple advantages such as high sensitivity, high precision, convenience, and easy operation, making UPT-POCT test an excellent diagnostic tool for emergency diseases, such as cardiovascular circulatory diseases, embolic diseases and infectious diseases. This chapter will discuss the advantages and applications of the UPT-POCT technology in emergency medicine.

Keywords UPT-POCT · Emergency diagnosis · Circulatory system diseases · Blood embolism · Infectious diseases

7.1 Emergency Medicine

The clinical emergency department is often faced with severe and acute diseases, such as cardiovascular and cerebrovascular diseases, embolic diseases and acute infectious diseases. Once the treatment is not timely, the patient's health will be seriously damaged. The clinician makes timely diagnosis and treatment for the patient mainly according to the patient's vital signs, medical history, external manifestations, and corresponding medical tests. Emergency medicine requires quick and accurate clinical test results in the shortest possible time, thus providing clinicians with timely and reliable information for the diagnosis and treatment. Therefore, diagnostics with high performances (efficiency, convenience, speed and accuracy) are particularly

Y. Li · R. Xiao · Y. Zhao (✉)
Beijing Key Laboratory of POCT for Bioemergency and Clinic (No. BZ0329), Beijing 100071, People's Republic of China
e-mail: zhaoyong179@139.com

Y. Li · R. Xiao
Beijing Hotgen Biotechnology Co., Ltd., Beijing Z-Park Daxing Biomedical Industry Base, No. 9, Tianfujie, Beijing 102600, China

Y. Zhao
Beijing Institute of Microbiology and Epidemiology, Academy of Military Medical Sciences, No. 20, Dongdajie, Fengtai District, Beijing 100071, China

R. Yang (ed.), *Principles and Applications of Up-converting Phosphor Technology*,
https://doi.org/10.1007/978-981-32-9279-6_7

important for the emergency medical. At present, there are two major development directions for emergency diagnosis research.

One direction is using large, automated and streamlined operating equipments to realize high-throughput, pipeline-based clinical testing; and make full use of the medical laboratory's information management system (such as the LIS system) for data transmission and diagnostic analysis. This model can greatly improve the accuracy and precision of clinical tests, however, it is highly dependent on the central laboratory, and it is not suitable for bedside diagnosis in emergency conditions.

Another important development direction of emergency laboratory medicine is poin-of-care testing (POCT), which can be carried out outside the laboratory and close to the patients, using portable and movable instruments and kits, and can report the test results quickly and timely. And it is not limited by the testing site and testing time of the central laboratory, do not require professionally trained personnel, which can better meet the needs of emergency medical (Di Serio et al. 2005).

With the development of biological sciences, laboratory medicine, immunology and new technology, POCT techniques and products have developed into three stages, the first stage (qualitative testing via color change), the second stage (semi-quantitative testing via grayscale detection) and the third stage (accurate quantitative testing via optical detection). The emergence of precision quantitative testing technique has greatly improved the sensitivity and accuracy of POCT detections, making POCT technique an indispensable tool for emergency medical diagnosis and treatment (Aachmann-Andersen et al. 2012; Di Serio et al. 2003; Zydron et al. 2011).

7.2 Emergency Diagnosis

7.2.1 Emergency Diagnosis of Circulatory System Diseases

Circulatory system diseases in the emergency department are mostly acute coronary syndrome (ACS) and acute myocardial infarction (AMI). The role of myocardial biomarker detection in the diagnosis of ACS and AMI has been repeatedly emphasized in relevant guidelines published by the College of Cardiology in the United States, Europe, and China. Among the biomarkers, troponin I (cTnI), troponin T (cTnT) and creatine kinase isoenzyme (CK-MB) are important serological diagnostic biomarkers for myocardial injury (Kehl et al. 2012); myoglobin (MYO) is an important serological biomarker for early diagnosis of AMI. In addition, cardiac fatty acid binding protein (H-FABP) (Tanaka 2006), N-terminal B-type natriuretic peptide precursor (NT-proBNP) (Januzzi et al. 2006), and human plasma lipoprotein phospholipase A2 (PLA2) (Macphee et al. 2005) are also effective biomarkers for early diagnosis of cardiovascular circulatory diseases.

The rapid and accurate detection of these cardiovascular disease serological biomarkers is very important for early identification of ACS and AMI. It was suggested that clinical samples of myocardial injury biomarkers should be tested immediately after sample collection, and the turnaround time of the test should be less than 1 h. It is also recommended that clinicians, physicians, nurses, as well as laboratory physician should be familiar with the use of POCT instruments and products.

7.2.2 *Emergency Diagnosis of Blood Embolism Diseases*

Blood embolism is a disease caused by thrombosis in blood vessels. It has a high incidence rate and ranks first in the global total disease mortality rate, and the incidence rate is increasing year by year. Thromboembolic diseases mainly involve the brain, heart, lungs, and peripheral vascular system, which seriously threaten human life and health. How to reduce morbidity and mortality of thromboembolic diseases is one of the focus and hotspots of modern emergency medical research.

A large number of studies have found that D-dimer in plasma is a specific indicator of the response to fibrinolytic function (Wells et al. 2004). D-dimer is a degradation product of fibrin after cross-linking with activated factor XIII and then by plasmin. Therefore, D-dimer will increase as long as there is activated thrombus formation or fibrinolysis in the blood vessels. D-dimer is recognized as one of the biomarkers of fibrinolytic response. It is recommended as a diagnostic biomarker for emergency pulmonary embolism and venous thrombosis.

7.2.3 *Emergency Diagnosis of Infectious Diseases*

Infectious diseases are bacterial or viral infections that usually occur in people with impaired autoimmune function or decreased resistance. Most infectious diseases can be cured as soon as possible if the diagnosis is timely and the intervention is appropriate. If it is not diagnosed and treated in time, it will develop to be serious life-threatening, especially for infants and elderly patients.

At present, commonly used infectious disease biomarkers include procalcitonin (PCT), C-reactive protein (CRP), interleukin 6 (IL-6), serum amyloid A (SAA) and the like (Goulart et al. 2010). Individual detection or combined detection of the biomarkers can effectively and early identify the infectious diseases.

7.3 UPT-POCT Advantages in Emergency Diagnosis

UPT-POCT technology is a representative POCT platform for quantitative detections and diagnosis. It has the advantages of high detection sensitivity, easy operation and rapid detection speed. These advantages make UPT-POCT an excellent diagnostic tool in emergency medical. After years of development, UPT-POCT technology has been successfully applied to the rapid diagnosis of cardiovascular circulatory diseases, embolic diseases and infectious diseases. The application of UPT-POCT technology to emergency testing has the following advantages:

(1) Fast detection speed

The UPT-POCT test method can obtain quantitative test results in just 15 min, which can provide clinicians with timely and reliable diagnosis and treatment basis.

(2) Comprehensive testing items

There are a variety of UPT-POCT diagnostic kits for cardiovascular diseases, thrombotic diseases and acute infections, including cTnI, CK-MB, MYO, NT-proBNP, H-FABP, Lp-PLA2, D-dimer, CRP, IL-6, and SAA.

(3) Applicable to various clinical samples

The UPT-POCT test kit is suitable for human serum, plasma and whole blood samples, requiring only 50–100 μl sample.

(4) Simple detection methods and equipment

Detection methods of the UPT-POCT diagnostic kits for different items are basically the same. Users only need to master one set of processing procedures to complete various detections. The basic operation steps are described as follows:

- Place the test card on a flat surface;
- Take 100 μl of the sample into the card;
- Reaction at room temperature for 15 min;
- Insert the test card into the biosensor for quantitative test results.

(5) Transmittable test results

It is available to directly transmit the test results to the laboratory information system, in order to meet the requirements of emergency medicine management.

7.4 Application of UPT-POCT in Emergency Medicine

After years of development, there have been a variety of UPT-POCT test kits for clinical diagnosis of cardiovascular diseases, thrombotic diseases and acute infections, including test kits for cTnI, CK-MB, MYO, NT-proBNP, H-FABP, Lp-PLA2, D-dimer, PCT, CRP, and IL-6 etc. These kits have been certificated and approved by the China Food and Drug Administration (CFDA) and have been successfully applied to emergency medicine in clinical hospitals with well performances (Table 7.1).

As a new type of POCT technology, UPT-POCT technology is portable, fast and easy to use. Compared with the common large-scale instrument detection method (magnetic particle chemiluminescence method), the test results of the above cardiovascular disease diagnostic biomarkers, blood embolic diseases and infection biomarkers showed no significant difference, through a large number of clinical sample evaluation experiments.

Table 7.1 UPT-POCT kits for emergency medicine

No.	Test item	Limit of detection	Quantification range
1	cTnI	0.1 ng/ml	0.1–40 ng/ml
2	CK-MB	2 ng/ml	2–500 ng/ml
3	MYO	5 ng/ml	5–1000 ng/ml
4	H-FABP	1 ng/ml	2–100 ng/ml
5	NT-proBNP	5 pg/ml	5–35000 pg/ml
6	Lp-PLA2	5 ng/ml	5–800 ng/ml
7	D-dimer	25 ng/ml	50–2500 ng/ml
8	PCT	0.02 ng/ml	0.02–50 ng/ml
9	CRP	0.5 ng/ml	0.5–150 ng/ml
10	IL-6	4 pg/ml	4–4000 pg/ml

7.5 Summary

In summery, UPT-POCT is a representative technology platform of the quantitative detection techniques. It has multiple advantages such as high sensitivity, high precision, convenience, and easy operation, making UPT-POCT an indispensable part of emergency medicine. With the further development of UPT-POCT technology, it will be able to rapidly promote the application of POCT in the emergency department and better support the emergency medicine.

References

Aachmann-Andersen NJ, Bjerrum PJ, Rasmussen SW, Schmidt TA. POCT is a true asset in the emergency department. Scand J Trauma, Resusc Emerg Med. 2012;19(S2).

Di Serio F, Antonelli G, Trerotoli P, Tampoia M, Matarrese A, Pansini N. Appropriateness of point-of-care testing (POCT) in an emergency department. Clin Chim Acta. 2003;333(2):185–9.

Di Serio F, Amodio G, Varraso L, Campaniello M, Coluccia P, Trerotoli P, Antonelli G, Pansini N. Integration between point-of-care cardiac markers in an emergency/cardiology department and the central laboratory: methodological and preliminary clinical evaluation. Clin Chem Lab Med. 2005;43(2):202–9.

Goulart LR, Vieira CU, Freschi APP, Capparelli FE, Fujimura PT, Almeida JF, Ferreira LF, Goulart IMB, Brito-Madurro AG, Madurro JM. Biomarkers for serum diagnosis of infectious diseases and their potential application in novel sensor platforms. Crit Rev™ Immunol. 2010;30(2):201–22.

Januzzi JL, van Kimmenade R, Lainchbury J, Bayes-Genis A, Ordonez-Llanos J, Santalo-Bel M, Pinto YM, Richards M. NT-proBNP testing for diagnosis and short-term prognosis in acute destabilized heart failure: an international pooled analysis of 1256 patients: the international collaborative of NT-proBNP study. Eur Heart J. 2006;27(3):330–7.

Kehl DW, Iqbal N, Fard A, Kipper BA, De La Parra LA, Maisel AS. Biomarkers in acute myocardial injury. Transl Res J Lab Clin Med. 2012;159(4):252–64.

Macphee CH, Nelson JJ, Zalewski A. Lipoprotein-associated phospholipase A2 as a target of therapy. Curr Opin Lipidol. 2005;16(4):442–6.

Tanaka T, Sohmiya Ki, Kitaura Y, Takeshita H, Morita H, Ohkaru Y, Asayama K, Kimura H: Clinical evaluation of point-of-care-testing of heart-type fatty acid-binding protein (H-FABP) for the diagnosis of acute myocardial infarction. J Immunoass Immunochem. 2006;27(3):225–38.

Wells PS, Anderson DR, Rodger M. Evaluation of D-Dimer in the diagnosis of suspected deep-vein thrombosis. ACC Curr J Rev. 2004;13(1):15.

Zydron CT, Woodworth A, Storrow AB. The future of point-of-care testing in emergency departments. Exp Opin Med Diagn. 2011;5(3):175–81.

Chapter 8
Application of UPT-POCT in Internal Medicine

Yanzhao Li, Honggang Zhang and Yong Zhao

Abstract The high medical cost, the trend of individualized medical care and the huge demand for rapid diagnosis are the main factors driving the rapid development of UPT-POCT technology in the field of internal medicine. UPT-POCT technology, as an early diagnostic tool with high sensitivity and convenience, can not only serve as an important supplement to the central laboratory, but also play an important role in primary clinical institutions. This chapter will discuss the application of UPT-POCT technology in internal medicine diseases, including liver diseases, cardiovascular diseases and endocrine-related diseases.

Keywords UPT-POCT · Liver disease · Cardiovascular disease · Endocrine disease

8.1 Internal Medicine

As a comprehensive clinical discipline, internal medicine mainly studies diseases with systemic, complex, and chronic characteristics, such as liver related diseases, cardiovascular and cerebrovascular diseases and endocrine related diseases. In current clinical practice, clinicians need to obtain information on all aspects of the patient, including the medical history, symptoms, medical imaging examination, and test results from the central laboratory. For diseases that require long-term treatment, clinicians need to continuously monitor changes in the condition to identify disease progression and determine the best treatment.

Y. Li · H. Zhang · Y. Zhao (✉)
Beijing Key Laboratory of POCT for Bioemergency and Clinic (No. BZ0329), Beijing 100071, People's Republic of China
e-mail: zhaoyong179@139.com

Y. Li · H. Zhang
Beijing Hotgen Biotechnology Co., Ltd., Beijing Z-Park Daxing Biomedical Industry Base, No. 9, Tianfujie, Beijing 102600, China

Y. Zhao
Beijing Institute of Microbiology and Epidemiology, Academy of Military Medical Sciences, No. 20 Dongdajie, Fengtai District, Beijing 100071, China

R. Yang (ed.), *Principles and Applications of Up-converting Phosphor Technology*,
https://doi.org/10.1007/978-981-32-9279-6_8

With the implementation and promotion of the grading diagnosis and treatment system in China, primary care clinics are facing tremendous pressure and challenges. UPT-POCT technology, as a representative third-generation POCT technology, can not only serve as an important supplement to the central laboratory, but also play an important role in primary clinics (Turner et al. 2016). In addition, the high medical cost, the trend of individualized medical care and the huge demand for rapid diagnosis are the main factors driving the rapid application of UPT-POCT technology in primary clinics.

8.2 The Application of UPT-POCT in Liver Diseases

There are about 100 million hepatitis patients in China, including more than 30 million people at high risk of liver cancer. More than two-thirds of patients of liver cancer are found in terminal stage, partly because the incubation period of liver cancer is longer and the symptoms are not obvious, at present. Therefore, early screening and diagnosis of liver fibrosis and cirrhosis is important for early identification of liver cancer and reduction of the incidence.

In clinical practice, liver biopsy has been used as the "gold standard" for the diagnosis of liver diseases (Pattullo et al. 2007). However, liver biopsy is traumatic, causing great pain to patients, and it is difficult to perform multiple tests in a short period of time (Grant and Neuberger 1999). Therefore, it can not be widely used as a routine screening method for liver fibrosis and cirrhosis. In the clinical diagnosis and treatment of liver diseases, finding a more efficient and convenient detection method that can replace liver biopsy has been a research hotspot.

Based on the currently known serological diagnostic biomarkers for liver fibrosis (Sebastiani 2012; Cequera 2014), five UPT-POCT assay kits for liver fibrosis have been successfully developed, including serum hyaluronic acid (HA) and laminin (LN), type IV collagen (CIV), type III procollagen amino terminal peptide (PIIINP) and serum tissue inhibitor of metalloproteinase-1 (TIMP-1) assay kit. These kits enable the early screening and diagnosis of liver fibrosis and cirrhosis in high-risk populations of liver cancer, effectively avoiding the shortcomings of liver biopsy. In addition, a large number of studies have shown that Golgi protein 73 (GP73) is a very sensitive biomarker of liver inflammation and liver fibrosis (Gu et al. 2009; Tian et al. 2011). The first GP73 assay kit was successfully developed based on the UPT-POCT technology, which can further demonstrate the clinical value of GP73 in the diagnosis of liver diseases.

For end-stage liver cancer, UPT-POCT kits for alpha-fetoprotein (AFP) and abnormal prothrombin (DCP) have been successfully developed, which can be used for dynamic monitoring of patients with confirmed liver cancer.

The above-mentioned UPT-POCT assay kits, including GP73, HA, LN, CIV, PIIINP, TIMP-1, AFP and DCP assay kits, have been approved by the China Food and Drug Administration (CFDA) and widely applied in the clinical diagnosis and treatment.

8.3 The Application of UPT-POCT in Circulatory System Diseases

Cardiovascular and cerebral circulatory diseases usually develop rapidly if they are not recognized timely in the early stage. A variety of UPT-POCT detection kits have been developed for the serological biomarkers of cardiovascular and cerebrovascular diseases, including troponin I (cTnI), troponin T (cTnT), and creatine kinase isoenzyme (CK-MB), myoglobin (MYO), cardiac fatty acid binding protein (H-FABP), N-terminal B-type natriuretic peptide precursor (NT-proBNP), human plasma lipoprotein phospholipase A2 (PLA2), and D-dimer etc. These kits can be used for early diagnosis and identification of cardiovascular and cerebrovascular diseases.

8.4 The Application of UPT-POCT in Endocrine System Diseases

Endocrine system diseases generally refer to the diseases caused by endocrine disorders. Among them, female polycystic ovary syndrome (PCOS) is the most common endocrine disorder disease in women of childbearing age, which can cause anovulatory infertility and seriously endanger women's physical and mental health (Durlinger 2002; Goodarzi et al. 2011).

Abnormal anti-Mullerian hormone (AMH) is considered to be the most stable and accurate serological biomarker for the PCOS (van Rooij et al. 2005). At present, the methods for detecting AMH levels in blood mainly include enzyme-linked immunosorbent assay (ELISA), chemiluminescence assay (CLIA) and UPT-POCT assay. Though the performance of UPT-POCT assay is slightly lower than that of the CLIA, it could better meet the needs of clinical diagnosis because it is more rapid, convenient, easy to operate, and do not require large-scale detection equipment. The main parameters of the UPT-POCT assay kit and the CLIA-based assay kit are shown in Table 8.1.

Table 8.1 Comparisons between UPT-POCT kit and CLIA kit for AMH detection

Main parameters	UPT-POCT assay kit	CLIA-based assay kit
Limit of detection	0.15 ng/ml	0.05 ng/ml
Quantification range	0.15–16 ng/ml	0.05–23 ng/ml
Precision	≤15%	≤15%
Specificity	No cross-reaction with high concentrations of bilirubin, hemoglobin, triglyceride, inhibin A, activin A, luteinizing hormone, and follicle stimulating hormone	No cross-reaction with high concentrations of bilirubin, hemoglobin, triglyceride, inhibin A, activin A, luteinizing hormone, and follicle stimulating hormone
Device requirement	Small portable instrument	Large instrument

In addition, UPT-POCT can also be applied to the rapid diagnosis of other internal diseases such as rheumatoid, kidney injury, and premature birth prediction. For example, the anti-cyclic citrullinated peptide antibody (anti-CCO) assay kit for rheumatoid detection, the neutrophil gelatinase-associated lipocalin (NGAL) assay kit for renal injury detection, the fetal fibronectin (fFN) assay kits for preterm birth prediction, etc. These kits have all been approved by the CFSA and were successfully applied in clinical applications.

In short, UPT-POCT technology, as an early diagnostic tool with high sensitivity and convenience, is an important supplement to the central laboratory, providing efficient and accurate support for medical examination, diagnosis, intervention, and prognosis of internal diseases. And, it is easy to promote and apply the UPT-POCT technology in primary medical institutions, greatly helping to improve the level of primary medical care.

References

Cequera A, García de León Méndez MC. Biomarkers for liver fibrosis: advances, advantages and disadvantages. Revista de Gastroenterología de México (English Edition). 2014;79(3):187–99.

Durlinger A. Regulation of ovarian function: the role of anti-Mullerian hormone. Reproduction. 2002;124(5):601–9.

Goodarzi MO, Dumesic DA, Chazenbalk G, Azziz R. Polycystic ovary syndrome: etiology, pathogenesis and diagnosis. Nat Rev Endocrinol. 2011;7(4):219–31.

Grant A, Neuberger J. Guidelines on the use of liver biopsy in clinical practice. Gut. 1999;45(Supplement 4):iv1–iv11.

Gu Y, Chen W, Zhao Y, Chen L, Peng T. Quantitative analysis of elevated serum Golgi protein-73 expression in patients with liver diseases. Ann Clin Biochem. 2009;46(Pt 1):38–43.

Pattullo V, Tomlinson J, Bell C. Education and imaging. Gastrointestinal: solitary rectal ulcer syndrome. J Gastroenterol Hepatol. 2007;22(12):2362.

Sebastiani G. Serum biomarkers for the non-invasive diagnosis of liver fibrosis: the importance of being validated. Clin Chem Lab Med. 2012;50(4):595–7.

Tian L, Wang Y, Xu D, Gui J, Jia X, Tong H, Wen X, Dong Z, Tian Y. Serological AFP/Golgi protein 73 could be a new diagnostic parameter of hepatic diseases. Int J Cancer. 2011;129(8):1923–31.

Turner PJ, Van den Bruel A, Jones CH, Pluddemann A, Heneghan C, Thompson MJ, Price CP, Howick J. Point-of-care testing in UK primary care: a survey to establish clinical needs. Fam Pract. 2016;33(4):388–94.

van Rooij IA, Broekmans FJ, Scheffer GJ, Looman CW, Habbema JD, de Jong FH, Fauser BJ, Themmen AP, te Velde ER. Serum antimullerian hormone levels best reflect the reproductive decline with age in normal women with proven fertility: a longitudinal study. Fertil Steril. 2005;83(4):979–87.

Chapter 9
Application of UPT-POCT in Combat-Related Traumatic Infection

Yanzhao Li, Ruifeng Xiao and Yong Zhao

Abstract Traumatic infection is a common disease for military man or the wounded civilians. If not timely treated, it could cause serious secondary sepsis and other complications. In order to achieve rapid diagnosis and treatment of traumatic infections, it calls for small portable devices, rapid diagnostic kits, and effective solution that can be used in the battlefield or other complex field. This chapter will mainly introduce the UPT-POCT based-portable biosensors and diagnostic kits, and their applications in combat-related traumatic infection.

Keywords Traumatic infection · Sepsis · UPT-POCT assay · Biosensor

9.1 Introduction

Traumatic infection is a common disease in the battlefield environment. If it is not treated in time, it can cause sepsis, septic shock, multiple organ dysfunction syndrome (MODS), and even death (Tribble et al. 2011). In addition, the battlefield environment is complex, such as a large area, urgent treatment time, and other uncertainty factors. In this special environment, rapid diagnosis and treatment is essential to improve the treatment of traumatic infection and secondary sepsis. In addition, in non-military operations, such as earthquakes, floods, fires, explosions, and accidental traffic accidents, the rapid and accurate diagnosis of traumatic infections and secondary sepsis is also important for the wounded.

Y. Li · R. Xiao · Y. Zhao (✉)
Beijing Key Laboratory of POCT for Bioemergency and Clinic (No. BZ0329), Beijing 100071, People's Republic of China
e-mail: zhaoyong179@139.com

Y. Li · R. Xiao
Beijing Hotgen Biotechnology Co., Ltd., Beijing Z-Park Daxing Biomedical Industry Base, No. 9, Tianfujie, Beijing 102600, China

Y. Zhao
Beijing Institute of Microbiology and Epidemiology, Academy of Military Medical Sciences, No. 20, Dongdajie, Fengtai District, Beijing 100071, China

R. Yang (ed.), *Principles and Applications of Up-converting Phosphor Technology*,
https://doi.org/10.1007/978-981-32-9279-6_9

In recent years, many country militaries have attached great importance to early-stage emergency treatment research, in order to enable casualties of light or moderate disease to quickly recover their combat capabilities (O'Brien et al. 2008). For this purpose, military medical treatment should meet the requirements of rapid treatment, easy delivery and convenient operation, which puts higher requirements on the POCT technology.

9.2 The Diagnostics of Combat-Related Traumatic Infection

The traditional clinical diagnosis of traumatic infection and secondary sepsis relies mainly on microscopic examination and microbial culture. Microscopic examination can be used for pathogenic bacteria with obvious characteristics in morphology and staining. The culture method usually takes 48–72 h (Xafranski et al. 2013), which often delays the diagnosis and treatment and even leads to the death of patients with sepsis. The current biochemical method using large-scale equipment cannot meet the complex environment in the battlefield. Therefore, small portable devices and diagnostic reagents that can adapt to the battlefield environment have become research hotspots.

9.3 The Application of UPT-POCT in Combat-Related Traumatic Infection

At present, portable UPT-POCT biosensors and a variety of diagnostic kits have been successfully developed, including procalcitonin (PCT) assay kits, interleukin-6 (IL-6) assay kits, and C-reactive protein (CRP) assay kits. These products provide a fast and effective solution for the early diagnosis of traumatic infection and secondary sepsis.

9.3.1 Portable UPT-POCT Biosensor

There are two main types of UPT biosensor, one is the desktop UPT-3A biosensor that can be used clinically, and the other is the handheld mini-UPT biosensor. Both can be used for trauma infection diagnosis and treatment in the battlefield environment (Fig. 9.1). The UPT-3A biosensor can be carried out in the vehicle or in the temporary field hospital. The handheld mini-UPT biosensor has a smaller size and weight, equipped with a battery that can stand for at least 4 h, which is especially suitable for the early detection of traumatic infection in the wild.

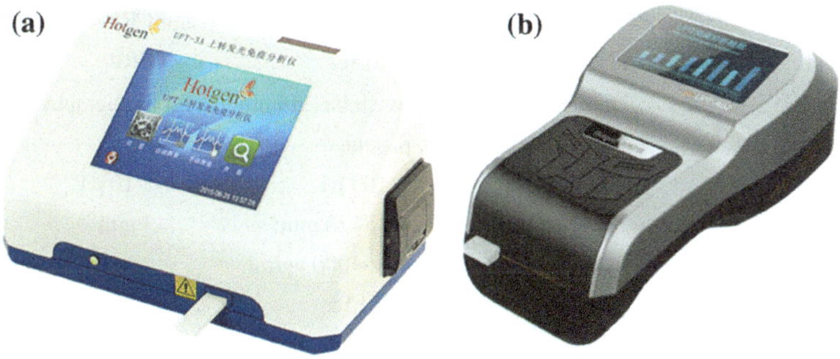

Fig. 9.1 The UPT-3A biosensor (**a**) and the handheld mini-UPT biosensor (**b**)

The UPT-3A biosensor and the mini-UPT biosensor have the same detection performances; both are easy to operate and perform well. The matching PCT, IL-6 and CRP test kits can be carried or stored in a portable transport case, which is convenient for hygienists to conduct timely diagnosis and treatment of trauma patients.

9.3.2 UPT-POCT Diagnostic Kits for Combat-Related Traumatic Infection and Sepsis

UPT-POCT technology has great advantages in the detection of traumatic infection and sepsis. (1) Fast detection, high sensitivity, and accurate quantification. (2) High specificity, strong anti-interference ability and strong stability. (3) It can directly detect samples of different types, including soil, powder, animal organs and body fluid samples (Zhang et al. 2014; Li et al. 2009). (4) It is easy to operate and does not require complicated instruments, suitable for the field detections. (5) It can be carried by person or applied in a vehicle or a tent hospital.

At present, PCT, IL-6 and CRP detection kits based on UPT-POCT technology platform have been successfully developed and widely applied in clinical practice. The main parameters are shown in Table 9.1.

Numerous studies have shown that it is not reliable enough to diagnose a disease by the change of one biomarker. The diagnosis should be made by combining the patient's clinical manifestations and the test results of multiple infectious indicators, which can provide clinicians with more accurate diagnostic information. The serum biomarkers of IL-6, PCT and CRP can be combined for the diagnosis of infection and sepsis (Plesko et al. 2016; Gao et al. 2017). Among them, the IL-6 level in serum is elevated before the appearance of obvious symptoms of sepsis, which is an early indicator of sepsis (Damas et al. 1992). The concentration peak of serum PCT peak appears after 24 h of sepsis, and it has a longer half-life, which is a specific diagnostic indicator of infectious sepsis (Azevedo et al. 2012). CRP can be non-

Table 9.1 Main parameters of UPT-POCT detection kits for PCT, IL-6 and CRP

Main parameters	PCT	IL-6	CRP
Principle	Double-antibody-sandwich based immunochromatography		
Sample type	Whole blood, serum, plasma		
Sample volume	100 μl	100 μl	5-10 μl
Detection time	15–20 min	15–20 min;	3 min
Quantitative range	0.02–50 ng/ml	4–4000 pg/ml	0.5–150 ng/ml
Precision	≤15%	≤15%	≤15%
Specificity	No cross-reactivity with high concentrations of bilirubin and HAS		
Application	Diagnostic indicators of bacterial infection and sepsis		

specifically elevated and can be used as an auxiliary diagnostic indicator for infection (Su et al. 2013). The combination of the three indicators can be used for continuous and dynamic monitoring of high-risk populations of sepsis, with important value for early detection and treatment of sepsis.

In summary, UPT-POCT technology can achieve timely and efficient diagnosis of traumatic infection and secondary sepsis, which is of great significance for improving the cure rate of traumatic infection and reducing the military combat loss caused by sepsis. With the further development and improvement of UPT technology, it will play a more important role in military medicine.

References

Azevedo JD, Torres O, Malafaia O. Procalcitonin as a prognostic biomarker of severe sepsis and septic shock. Revista Do Colégio Brasileiro De Cirurgiões. 2012;39(6):456–461.

Damas P, Ledoux DI, Nys M, Vrindts Y, De Groote D, Franchimont P, Lamy M. Cytokine serum level during severe sepsis in human IL-6 as a marker of severity. Ann Surg. 1992; 215(4):356.

Gao L, Liu X, Zhang D, Xu F, Chen Q, Hong Y, Feng G, Shi Q, Yang B, Xu L. Early diagnosis of bacterial infection in patients with septicopyemia by laboratory analysis of PCT, CRP and IL-6. Exp Ther Med. 2017;13(6):3479–3483.

Li L, Zhou L, Yu Y, Zhu Z, Lin C, Lu C, Yang R. Development of up-converting phosphor technology-based lateral-flow assay for rapidly quantitative detection of hepatitis B surface antibody. Diagn Microbiol Infect Dis. 2009;63(2):165–172.

O'Brien K, Cadbury N, Rollnick S, Wood F. Sickness certification in the general practice consultation: the patients' perspective, a qualitative study. Fam Pract. 2008;25(1):20–26.

Plesko M, Suvada J, Makohusova M, Waczulikova I, Behulova D, Vasilenkova A, Vargova M, Stecova A, Kaiserova E, Kolenova A. The role of CRP, PCT, IL-6 and presepsin in early diagnosis of bacterial infectious complications in paediatric haemato-oncological patients. Neoplasma. 2016;63(5):752–760.

Su L, Feng L, Song Q, Kang H, Zhang X, Liang Z. Diagnostic value of dynamics serum sCD163, sTREM-1, PCT, and CRP in differentiating sepsis. Mediat Inflamm. 2013;2013(5):969875.

Tribble DR, Conger NG, Fraser S, Gleeson TD, Wilkins K, Antonille T, Weintrob A, Ganesan A, Gaskins LJ, Li P, et al. Infection-associated clinical outcomes in hospitalized medical

evacuees after traumatic injury: trauma infectious disease outcome study. J Trauma. 2011;71(1 Suppl):S33–S42.
Xafranski H, Melo AS, Machado AM, Briones MR, Colombo AL. A quick and low-cost PCR-based assay for Candida spp. identification in positive blood culture bottles. BMC Infect Dis. 2013;13:467.
Zhang P, Liu X, Wang C, Zhao Y, Hua F, Li C, Yang R, Zhou L. Evaluation of up-converting phosphor technology-based lateral flow strips for rapid detection of Bacillus anthracis Spore, Brucella spp., and Yersinia pestis. PLoS One. 2014;9(8):e105305.

Chapter 10
Application of UPT-POCT in Medical Relief for Disasters

Yanzhao Li, Miao Jia and Pingping Zhang

Abstract A variety of disasters with high frequency occur around the world. Shortage of equipment and professional, as well as interruptions of power and communication may occur for the disaster scene. Personnel who performs inspections may also in charge of hemospasia, blood transfusion, and equipment maintenance. Therefore, point of care testing (POCT) plays an indispensable role in disaster relief, and can provide a large number of irreplaceable experimental diagnostic data for the entire medical relief for disasters. UPT-POCT is portable, easy to operate, fast, accurate and reliable, as well as requires a trace amount of sample, and it can be used under extreme environmental conditions, meeting the various requirements for disaster scenes.

Keywords Medical relief · Earthquake · Flood · Conflagration and explosion · UPT-POCT

10.1 The Characteristic of Medical Relief for Disasters

10.1.1 The Present Situation of Medical Relief for Disasters in the World

The World Health Organization (WHO) defines disasters as: disaster occurs when the destructive power of any event exceeds the endurance capacity of the area and the

Y. Li · M. Jia · P. Zhang (✉)
Beijing Key Laboratory of POCT for Bioemergency and Clinic (No. BZ0329), Beijing 100071, China
e-mail: acedreams@126.com

Y. Li · M. Jia
Beijing Hotgen Biotechnology Co., Ltd., Beijing Z-Park Daxing Biomedical Industry Base, No. 9 Tiantujie, Beijing 102600, China

P. Zhang
Institute of Microbiology and Epidemiology, Academy of Military Medical Sciences, No. 20 Dongdajie, Fengtai District, Beijing 100071, China

R. Yang (ed.), *Principles and Applications of Up-converting Phosphor Technology*,
https://doi.org/10.1007/978-981-32-9279-6_10

assistance of other regions is demanded, while any event that can cause the destruction to the facility, economy, the health of the person and the conditions of social health services may be involved. Disasters include natural hazards of meteorology, geology, geomorphology, hydrology, biology and so forth, as well as man-made disasters such as conflagrations, explosions and terrorist attacks operated by human (Elamein et al. 2015). Medical relief for disasters mainly depends on militaries in many countries, including not only the onsite medical service for the wounded and sick, but also the monitoring of epidemiological pathogens from water, soils and population (Michaud et al. 2019).

Medical relief for disasters is a discipline that studies emergency medical service, disease prevention, and health care under various disasters. There are currently two models in the world, namely the Anglo-American model and the French-German model. The former provides out-of-hospital rescue by professionals and sends patients to the hospital emergency department for therapy, while the latter provides emergency medical service outside the hospital by medical personnel. China adopts a combination of on-site ambulance and distribution of patients to the hospital. Portable medical inspection equipment meets the requirements of modern medical relief for disasters.

10.1.2 The Characteristic of Detection for Medical Relief for Disasters

Due to the suddenness and uncertainty of disasters, a lot of detection tasks for medical relief for disasters should be performed by nonprofessionals with inferior facilities in a poor working environment (Goyet et al. 2018).

10.1.2.1 Large Amounts of Inspection Work

For medical relief for disasters on site, the inspection work provides basic experimental diagnosis data for the clinical inspection, biochemical detection, blood transfusion and infectious disease monitoring projects. The contents of projects are complicated, greatly increasing the difficulty of on-site testing.

10.1.2.2 High Requirements for the Comprehensive Quality of the Inspectors

Inspectors must be responsible for inspection, quarantine, haemospasia, blood transfusion, epidemiological monitoring, management and maintenance of reagents and equipments in medical relief for disasters on site. The quality including knowledge and skills of inspectors must be comprehensive.

10.1.2.3 High Requirements for Equipment and Reagents

Interruptions of traffic, power and communication, as well as poor conditions at the disaster scene, may affect the performances of equipments and reagents for medical testing.

10.2 The Requirement of Detection for Medical Relief for Disasters

The characteristics of medical relief for disasters are contingence, complex, and uncertain. On the basis of tests covering clinical examination, biochemistry, immunity, microbiology, and blood transfusion, etc., other medical tests should be added according to the season and the type of disaster, such as schistosomiasis, malaria, dengue fever, cholera (Siddique et al. 1995) and other endemic diseases in disaster areas. It is necessary to have a suitable combination of inspection items for comprehensive medical service for different kinds of disasters such as earthquakes, floods, conflagrations and leakage of toxic chemicals. Due to the isolation from outside and the paralysis of water, electricity, communication and transportation at the large-scale disaster scenes, large analytical instruments in central laboratories will not be able to reach the site, while mobile, portable instruments or test strips will become the first choice for emergency medical treatment. After the data delivery of POCT, necessary medical information based on comprehensive physical examination of the wounded will be provided to clinicians, improving the success rate of rescue. Therefore, the detection system for medical relief for disasters must be equipped with easy-to-carry, self-powered and stable instruments, as well as easy-preserved reagents, and it also should be easy to operate, fast in detection and robust in anti-interference to ensure rapidity of medical examination with high accuracy.

10.2.1 The Diseases and Corresponding Detection Items in Earthquake Disasters

The crush syndrome and damage of important organs caused by the earthquake are the main causes for deaths of the wounded, and symptoms are mainly injury, local sensory disturbance and acute renal failure (Missair et al. 2013). Neutrophil gelatinase-associated lipocalin (NGAL) is one of the most effective serological markers for early diagnosis of acute kidney injury, because it increased faster than other molecules markers of kidney injury molecules (Noto et al. 2013). In addition to crush syndrome, there are other traumas such as fractures, and shocks, damage of skeletal and cardiac muscle, while the rapid detection of myocardial markers is indispensable.

What's more, infections often occur because the serious pollution for the environment and the poor facilities for the rescue of the wounded in earthquakes. The detection markers for infectious diseases are effective for early and accurate diagnosis of systemic sepsis, bacterial and viral infections, including procalcitonin (PCT), C-reactive protein (CRP) and interleukin-6 (IL-6).

The environmental barrier for prevention of disease spreads were weakened because of ecological damage and the deteriorated environment caused by earthquakes. For environmental problems, waste water, sewage and the bodies of victims can cause the growth of bacteria, mold, mosquitoes and parasites. A damaged local water supply system can increase the incidence of intestinal infectious diseases, as well as water-borne infectious diseases such as cholera, schistosomiasis and leptospirosis. Therefore, infectious disease tests should be carried out in earthquake disasters, and surveillance of serious infectious diseases must be done in natural foci such as anthrax, brucellosis and plague.

10.2.2 The Diseases and Corresponding Detection Items in Flood Disasters

China is one of the countries with frequent floods in the world. Floods can cause drowning, hypothermia and many other injuries. The concentrated placement of the victims will increase the possibility of outbreaks of infectious diseases. Studies have shown that the incidences of typhoid fever, paratyphoid fever, hepatitis, gastroenteritis and measles have increased after floods (Llewellyn 2006), therefore the diseases with the clinical symptom of fever and diarrhea (Yip and Sharp 1991) should be concerned. In addition, pollution of drinking water sources often occurs after flood disasters, therefore relevant foodborne pathogens must be quarantined.

10.2.3 The Diseases and Corresponding Detection Items in Conflagration and Explosion Disasters

After the conflagration, blood gas analysis and electrolyte analysis are needed due to blood transfusion and infusion. Infection and sepsis are common complications of burns, and detections of common pathogens can help monitor the condition. Explosion disasters can cause important visceral and mechanical damages, infection will develop without treatment in time. Therefore the detection markers related to the infectious disease are key testing items for conflagration and explosion disasters, such as PCT, CRP and IL-6.

10.3 UPT-POCT Detection Method for Medical Relief for Disasters

Combining UCP as bio-labels and classical immunochromatography as a rapid detection platform, the rapid quantitative detection of target antigens or antibodies can be realized based on photoelectric signals of UPT-POCT. Prior to the qualitative detection of traditional colloidal gold immunochromatography, UPT-POCT has the unique characteristic of stable luminescence, no interference background and high sensitivity, which is a representative technology platform for third-generation of POCT with precision and quantitation. UPT-POCT has the advantage of portability, operability and a small amount of sample demand, meeting the requirement of detection in extreme environmental conditions at the disaster scene. The information of products can be directly read by built-in chips or external barcodes and two-dimensional codes in the strip, and the results can be directly printed by the built-in printer of the series of UPT biosensor, refraining from the limitation of facilities at the disaster scene, what is the most important is that the rapid result acquisition within 20 min and reagent preservation at 4–30 °C are very practical and meet the requirement for detection at the disaster scene.

Please refer to Chap. 7 "Application of UPT-POCT in Emergency Medicine" and Chap. 8 "Application of UPT-POCT in Internal Medicine" for the information of UPT-POCT products for detection of myocardial markers, such as MYO, CK-MB, cTnI, FABP, NT-proBNP (Yang et al. 2017) and PLA2, as well as a marker for blood embolic disease, namely D-Dimer. In addition, please refer to Chapt. 9 "Application of UPT-POCT in Combat-Related Traumatic Infection" for the information of UPT-POCT products for detection of markers for infectious diseases, such as PCT, CRP and IL-6. The pathogens of common infections and intestinal infectious diseases in medical relief for disasters mainly include *V. cholerae* O139, *V. cholerae* O1 (Hao et al. 2017), *S. typhimurium*, *E. coli* O157, and *S. aureus* etc. (Zhao et al. 2016). The sensitivities of UPT-POCT for detections of food samples after enrichment and diarrhea samples is perfect, and please refer to Chap. 11 "Application of UPT-POCT for Detection of Foodborne Pathogens". If the disaster scenes overlap the natural foci for serious infectious diseases, the detection for these corresponding pathogens of UPT-POCT were described in Chap. 16 "Application of UPT-POCT in Anti-bioterrorism and Biosecurity".

10.4 Applications of UPT-POCT for Medical Relief for Disasters

At present, the equipments and reagents based on UPT-POCT method have been widely commercialized, and they can meet the needs of rapid inspections at the disaster scene because of the characteristics of small size, portability, simple operation, strong maneuverability and practicality of inspection items. UPT-POCT have been successfully applied to many fields, such as disease prevention and control, public security, firefighting, military, port, food and drug supervision, etc. The test products for diseases at the disaster scene are shown in Table 10.1.

In summary, the experience of dealing with emergencies and disaster events around the world over the years has proven the role and feasibility of POCT. POCT applications in medical relief for disasters provide scientific evidences for the decisions in disasters of corresponding plans, such as emergency management, first aid implementation, environmental management, and early warning start. As the representative technology of POCT, UPT-POCT is the first choice for first responders for medical inspections and managements in disasters. It meets the needs of on-site rapid detection and emergency disposal, that is, the suspicious samples with "unknown targets" can be quantitatively determined in the "shortest time" on site, improving the ability and standards of medical relief for disasters, and providing strong support for national biosafety, disaster relief, people's health, economic development and social stability.

Table 10.1 UPT-POCT products for diseases at the disaster scene

No.	Products
1	UPT-POCT detection kit for neutrophil gelatinase-associated lipocalin (NGAL)
2	UPT-POCT detection kit for cardiac troponin I (cTnI)
3	UPT-POCT detection kit for creatine kinase isoenzyme MB (CK-MB)
4	UPT-POCT detection kit for myoglobin (MYO)
5	UPT-POCT detection kit for myocardial fatty acid binding protein (H-FABP)
6	UPT-POCT detection kit for N-terminal prohormone of brain natriuretic peptide (NT-proBNP)
7	UPT-POCT detection kit for lipoprotein-associated phospholipase A2 (Lp-PLA2)
8	UPT-POCT detection kit for procalcitonin (PCT)
9	UPT-POCT detection kit for C reaction protein (CRP)
10	UPT-POCT detection kit for interleukin-6 (IL-6)
11	UPT-POCT detection kit for *V. cholerae* O139
12	UPT-POCT detection kit for *V. cholerae* O1
13	UPT-POCT detection kit for *S. typhi*
14	UPT-POCT detection kit for *E. coli* O157
15	UPT-POCT detection kit for *Y. pestis*
16	UPT-POCT detection kit for *S. aureus*
17	UPT-POCT detection kit for Norovirus

References

Elamein M, Bower H, Valderrama C, Zedan D, Rihawi H, Almilaji K, Abdelhafeez M, Tabbal N, Almhawish N, Maes S, et al. Attacks against health care in Syria, 2015–16: results from a real-time reporting tool. Lancet (London, England). 2017;390(10109):2278–2286.

Llewellyn M. Floods and tsunamis. Surg Clin N Am. 2006;86(3):557–578.

Goyet S, Rayamajhi R, Gyawali BN, Shrestha BR, Lohani GR, Adhikari D, Salvador E, Ofrin R, Vandelaer J, Samuel R. Post-earthquake health-service support, Nepal. Bull World Health Organization. 2018;96(4):286–291.

Hao M, Zhang P, Li B, Liu X, Zhao Y, Tan H, Sun C, Wang X, Wang X, Qiu H, et al. Development and evaluation of an up-converting phosphor technology-based lateral flow assay for the rapid, simultaneous detection of Vibrio cholerae serogroups O1 and O139. PLoS One. 2017;12(6):e0179937.

Michaud J, Moss K, Licina D, Waldman R, Kamradt-Scott A, Bartee M, Lim M, Williams J, Burkle F, Polyak CS, et al. Militaries and global health: peace, conflict, and disaster response. Lancet (London, England). 2019;393(10168):276–286.

Missair A, Pretto EA, Visan A, Lobo L, Paula F, Castillo-Pedraza C, Cooper L, Gebhard RE. A matter of life or limb? A review of traumatic injury patterns and anesthesia techniques for disaster relief after major earthquakes. Anesth Analg. 2013;117(4):934–941.

Noto A, Cibecchini F, Fanos V, Mussap M. NGAL and metabolomics: the single biomarker to reveal the metabolome alterations in kidney injury. Biomed Res Int. 2013;3:612032.

Siddique AK, Salam A, Islam MS, Akram K, Majumdar RN, Zaman K, Fronczak N, Laston S. Why treatment centres failed to prevent cholera deaths among Rwandan refugees in Goma, Zaire. Lancet (London, England). 1995;345(8946):359–361.

Yang X, Liu L, Hao Q, Zou D, Zhang X, Zhang L, Li H, Qiao Y, Zhao H, Zhou L. Development and evaluation of up-converting phosphor technology-based lateral flow assay for quantitative detection of NT-proBNP in blood. PLOS One. 2017;12(2):e0171376.

Yip R, Sharp TW. Acute malnutrition and high childhood mortality related to diarrhea. Lessons from the 1991 Kurdish refugee crisis. JAMA. 1993;270(5):587–590.

Zhao Y, Wang H, Zhang P, Sun C, Wang X, Wang X, Yang R, Wang C, Zhou L. Rapid multiplex detection of 10 foodborne pathogens with an up-converting phosphor technology-based 10-channel lateral flow assay. Sci Rep. 2016;6:21342.

Chapter 11
Application of UPT-POCT in Detection of Foodborne Pathogens

Yanzhao Li, Xingbo Ren, Hongrui Zhang and Pingping Zhang

Abstract Foodborne pathogens are responsible for most of diseases in food safety, mainly including *Salmonella*, *L. monocytogenes*, *E. coli* O157, *V. cholerae*, *V. parahaemolyticus*, *S. aureus* and *C. sakazakii*. Confirmation detection method is time-consuming, while rapid detection method is suitable to be used as basis for judicial and law enforcement on site. The requirements for limitations of pathogens in different types of food are varied, therefore the rapid quantitative detection method is desiderated. UPT-POCT can quantitatively detect dozens of foodborne pathogens after enrichment with high sensitivity and rapidity, and 17 commercial UPT-POCT detection kits have been developed and included in the reference list of equipments for food safety risk monitoring for provincial and municipal disease prevention and control institutions in China.

Keywords Foodborne pathogen · Limitation requirement · Confirmation detection · Rapid detection · UPT-POCT

11.1 Character for Detection for Foodborne Pathogens

Foodborne pathogens are one of the main factors affecting food quality and safety (Pereira et al. 2018), and "food safety" is the theme of World Health Day of the World Health Organization in 2015. According to incomplete statistics, there are as many as one billion foodborne diseases per year, of which more than 60% are caused by pathogenic bacteria. Each year about one million children die from diarrhea caused

Y. Li · X. Ren · H. Zhang · P. Zhang (✉)
Beijing Key Laboratory of POCT for Bioemergency and Clinic (no. BZ0329), Beijing 100071, China
e-mail: acedreams@126.com

Y. Li · X. Ren · H. Zhang
Beijing Hotgen Biotechnology Co., Ltd., Beijing Z-Park Daxing Biomedical Industry Base, No. 9 Tianfujie, Beijing 102600, China

P. Zhang
Beijing Institute of Microbiology and Epidemiology, Academy of Military Medical Sciences, No. 20 Dongdajie, Fengtai District, Beijing 100071, China

R. Yang (ed.), *Principles and Applications of Up-converting Phosphor Technology*,
https://doi.org/10.1007/978-981-32-9279-6_11

by foodborne pathogens. Many major accidents about public health were caused by foodborne pathogens, such as *E. coli* O157:H7 food poisoning incidents in Japan in 1996 and in China in 1999 involving tens of thousands of infections, the food safety incident related avian influenza in Southeast Asia in 2005, *L. monocytogenes*-contaminated melon incident and the epidemic of *salmonella* caused by live poultry that were extremely widespread in America in 2011. In China, the number of food poisoning caused by foodborne pathogens accounts for 40–60% of the total number of foodborne diseases.

11.1.1 The Main Foodborne Pathogens

The main pathogens for food safety are *Salmonella, L. monocytogenes, E. coli* O157, *V. cholerae, V. parahaemolyticus, S. aureus* and *C. sakazakii*. In addition, they also include various pathogenic bacteria such as *Brucella, C. botulinum, P. cocovenenans, C. jejuni* and *B. cereus*.

(1) *Salmonella* spp.
Salmonella spp., the causative agent of salmonellosis, is a Gram-negative Enterobacterium of Enterobacteriaceae. The main pathogens of *salmonella* spp. include *S. paratyphi* A, B, and C, *S. typhi, S. enteritidis* and *S. choleraesuis*. *Salmonella* can survive for 3–4 months at low temperature, which is the main pathogens for food poisoning in the world (Vinueza-Burgos et al. 2019). It is one of the health indicators for China's metrological certification/approval testing.

(2) *L. monocytogenes*
It is a Gram-positive Brevibacterium of genus Listeria, and facultative anaerobic. The clinical symptoms are mainly sepsis, meningitis and proliferation of mononuclear cells. Many types of foods can be contaminated by *L. monocytogenes*, such as aquatic products, dairy products, meat and vegetables.

(3) *E. coli* O157:H7
E. coli O157:H7 is a Gram-negative Brevibacterium with capsule, genus Escherichia, Enterobacteriaceae. *E. coli* O157:H7 can cause diarrhea, hemorrhagic enteritis, hemolytic uremic syndrome, purpura and other diseases, which are especially serious for children and the elderly and even lead to death. *E. coli* O157:H7 is mainly found in meat and vegetable products. Food safety incidents caused by consumption of beef and vegetable products contaminated with *E. coli* O157:H7 have occurred in many countries around the world.

(4) *S. aureus*
S. aureus is a spherical Gram-positive bacterium, genus Staphylococcus, irregularly arranged in a string of grapes for morphology. *S. aureus* can produce enterotoxin during growth and reproduction, which is destructive to the intestine, and it is also the most common pathogen in human purulent infection. *S. aureus* can be found in foods such as milk, meat, eggs, fish and their products,

and is also widely found in air, water, feed, soil, dust, as well as human and animal waste.

(5) *V. parahaemolyticus*

V. parahaemolyticus is a Gram-negative bacterium, genus Vibrio. It mainly pollutes aquatic products such as fish, shrimp, crab, shellfish, seaweed, or meat products by cross-contaminates. It is the main pathogen in coastal areas and some inland areas. *V. parahaemolyticus* has strong virulence to humans and animals, and the main clinical symptoms include the abdominal pain, vomiting, diarrhea and watery stools.

(6) *C. sakazakii*

C. sakazakii is a Gram-negative bacterium, Enterobacteriaceae. It is an important conditional pathogen in milk powder that can cause death in infants and young children (Lepuschitz et al. 2017). At the beginning of the 21st century, *C. sakazakii* was continuously detected from infant milk powders made by the international dairy giant company, and from then on public pay attention to the detection of this pathogen from dairy products. In 2005, China issued the Industry Standard for Detection Method of *C. sakazakii* in Milk Powder.

(7) *Shigella* spp.

Also as known as dysentery bacillus, *Shigella* spp. is a Gram-negative bacterium, the most common pathogen responsible for human bacterial dysentery. There are four groups for *Shigella* spp., including *S. dysenteriae*, *S. flexneri*, *S. boydii*, and *S. sonnei*. Dysentery is the most common intestinal infectious disease in summer and autumn, mainly transmitted by fecal oral route. The clinical symptoms are severe abdominal pain, diarrhea (watery stool with blood and mucus possibly) and fever. After years of continuous management and significant improvement of sanitary conditions, as well as surveillance for food safety, currently *Shigella* spp. has rarely been detected for food safety in China.

11.1.2 The Characteristics of Detection for Foodborne Pathogens

11.1.2.1 Classification for Detection

The detection of foodborne pathogens is a key mean for surveillance for food safety. At present, two detection methods are used for common pathogens for food safety in China. One is the confirmation inspection based on Food Microbiology Test (GB4789-2016) of National Standard for Food Safety, and the other is a rapid diagnostic method for detection of pathogens stated in the 120th item of Food Safety Law revised in 2015.

The confirmation detection method stated in National Standard GB4789-2016 mainly include culture of microorganism, as well as morphological, serological and biochemical identification, etc. The main operation steps include sample processing, pre-enrichment, separation, observation by microscope, biochemical experiments, serological typing (optional project), results report and other steps. The entire inspection period is at least 3–7 days, and the results are accurately quantified, rigorous and reliable. However, the disadvantages of that method are cumbersome, long time-consuming, and professional, which is not conducive to the control of potentially unsafe products (Rajapaksha et al. 2019).

Therefore, the rapid detection method for spot-check can be used for the food safety supervision and management work according to the Food Safety Law of the People's Republic of China. The preliminary screening method requires rapidity, simple operation, short time cost, and suitability for detection on-site. The results of rapid inspections can be used as a basis for the judicial and law enforcement in food safety for the food safety supervision department at or above the county level of the people's government in China.

11.1.2.2 Characteristic of Detection

The contamination of foodborne pathogens is varied with the changes of the environment and time (Zoellner et al. 2019), and food handler is also related to the prevalent of pathogens in food (Xu et al. 2019). The pathogens are unevenly distributed in the food, so that the detection results for the same sample at different times and different parts are different (Jagadeesan et al. 2019). The unsealed samples can't be used for repeat tests because of the risk of pollution. Therefore, instability of the contamination of pathogens is more serious than that of chemical pollution, and the requirements for timeliness and accuracy are very strict for detection of pathogens.

11.2 Requirement for Detection of Foodborne Pathogens

Requirements for limitations of foodborne pathogens were summarized by the General Standard for limitations of Pathogenic Bacteria in Food Safety (GB29921-2013), which was implemented on July 1, 2014, in China. The pathogens involved in the standard include *Salmonella* spp., *L. monocytogenes*, *E. coli* O157:H7, *S. aureus* and *V. parahaemolyticus*. The requirements for limitations of pathogens in different types of food are detailed in Table 11.1.

Stricter requirements for limitations of *E. coli*, *S. aureus* and *Salmonella* spp. were stated in "Infant Formula" (GB10765) and "Large Infant and Young Child Formula Foods" (GB 10767), the National Standard for Food Safety of China. The requirement for limitation of *C. sakazakii* was also added (Table 11.2).

Table 11.1 The requirements of limitations of foodborne pathogens in different types of food

Food types	Pathogens	Sampling method and limitation (mark by 25 g and 25 ml if not specially designated)			
		n	c	m	M
Meat product	*Salmonella* spp.	5	0	0	–
	L. monocytogenes	5	0	0	–
	S. aureus	5	1	100 CFU/g	1000 CFU/g
	E. coli O157:H7	5	0	0	–
Aquatic product	*Salmonella* spp.	5	0	0	–
	V. parahaemolyticus	5	1	100 MPN/g	1000 MPN/g
	S. aureus	5	1	100 CFU/g	1000 CFU/g
Instant egg product	*Salmonella* spp.	5	0	0	–
Cereal product	*Salmonella* spp.	5	0	0	–
	S. aureus	5	1	100 CFU/g	1000 CFU/g
Instant beans product	*Salmonella* spp.	5	0	0	–
	S. aureus	5	1	100 CFU/g	1000 CFU/g
Chocolate and cacao product	*Salmonella* spp.	5	0	0	–
Instant fruits and vegetables product (contain pickles)	*Salmonella* spp.	5	0	0	–
	S. aureus	5	1	100 CFU/g	1000 CFU/g
	E. coli O157:H7	5	0	0	–
Drink (except packaged drinking water and carbonated beverage)	*Salmonella* spp.	5	0	0	–
	S. aureus	5	1	100 CFU/g	1000 CFU/g
Frozen drinks	*Salmonella* spp.	5	0	0	–
	S. aureus	5	1	100 CFU/g	1000 CFU/g
Instant condiment	*Salmonella* spp.	5	0	0	–
	S. aureus	5	1	100 CFU/g	10000 CFU/g
	V. parahaemolyticus	5	1	100 MPN/g	1000 MPN/g
Nuts and seeds product	*Salmonella* spp.	5	0	0	–

Note n is the number of sample to be collected for one batch; C is maximum number of sample allowed to exceed n; m is the acceptable limit value for pathogen; M is maximum limit value for pathogen

Table 11.2 The requirements for limitations of foodborne pathogens for formula foods of infant and young child in China

Projects	Sampling scheme[a] and limitation (mark by CFU/g and CFU/mL if not specially designated)				Note
	n	c	m	M	
Total bacterial colonies[b]	5	2	1000	10000	
E. coli	5	2	10	100	
S. aureus	5	2	10	100	
C. sakazakii	3	0	0/100 g	–	0–6 months old infant food
Salmonella spp.	5	0	0/25 g	–	

Note a. The analysis and treatment of samples according to GB 4789; b. Not suitable for the products appended by active probiotics

11.3 UPT-POCT Detection for Foodborne Pathogens

11.3.1 The Performance of UPT-POCT for Detection of Foodborne Pathogens

The principle of UPT-POCT for detection of foodborne pathogens is mainly based on the double antibody sandwich method. Taking detection of *E. coli* as an example, UPT-POCT can detect *E. coli* O157:H7 with a concentration lower than 10^3 CFU/mL under the interference of 10^9 CFU/mL of other bacteria, and the coefficients of variation are all less than 10%.

In addition, multiple detection of foodborne pathogens can be performed. Simultaneous detection of ten foodborne pathogens can be complete at a time, including *S. paratyphi* A, B, and C, *V. cholerae* O1 and O139, *E. coli* O157, *S. enteritidis*, *S. choleraesuis*, *S. typhi*, and *V. parahaemolyticus*. The detection ranges for samples without enrichment are 10^5–10^9 CFU/mL, and 10 CFU/mL after enrichment (Zhao et al. 2016). Regression coefficient R^2 value for the quantitative curve exceeds 0.9, and the specificity is excellent for the other nine pathogens. As a whole, UPT-POCT can detect the food sample after enrichment, as well as the direct detection of diarrhea samples. After enrichment, the sensitivities meet the requirement of pathogens stated by the General Standard GB29921-2013.

11.3.2 The Advantage of UPT-POCT for Detection of Foodborne Pathogen

The UPT-POCT based on upconversion nanoparticles has the characteristics of high sensitivity, specificity and stability, zero background interference, and simple operation, compared with the rapid detection methods based on other nanomaterials, such as colloidal gold, latex particles, and fluorescent particles. (1) The detection method is universal. The applicability of UPT-LF is in favor of the development of not only new reagents for pathogen detection, but also the general solution for sample processing and the same method for sample treatment, leading to easy of detection for various pathogens with a set of treatment processes. In addition, a UPT-3A biosensor can be used for the detection of dozens of foodborne pathogens, therefore the funding and the placement space for equipment can be saved for the inspection organization. (2) The enrichment of UPT-POCT for foodborne pathogens is only 6–24 h, and the time of simple sample treatment and quantitative detection is less than 20 min, which is far shorter than 3–7 days required by international standard, making the law enforcement of food safety more timely and efficiently. (3) The detection results can be traceable. The detection results for UPT-POCT for common pathogenic microorganisms for food safety can be viewed immediately on the screen or printed by the built-in printer after the test completed. In addition, the results automatically stored by UPT biosensor can also be transferred to the computer for storage and analysis, so the traceability of the results is excellent.

11.4 Application

11.4.1 The Current Detection Method for Food Pathogen

There are many traditional detection methods stated by Food Microbiology Test (GB4789-2016) of National Standard for Food Safety, such as microbial culture, morphological identification and biochemical experiments. Many new methods and equipments have been developed for the detection of foodborne pathogens, including technology related improved counting, immunochromatographic detection, gene detection (Liu et al. 2019), microfluidic, biochip detection, biosensor detection, mass spectrometry detection, next-generation sequencing (Yang et al. 2019), and flow cytometry technology, etc. Among them, immunochromatography, microfluidic technology, and DNA amplification (Liu et al. 2019) are suitable for rapid screening and preliminary diagnosis for foodborne pathogens on site because of their rapidity, portability and simple operation. As a immunochromatography detection method, UPT-POCT is fast, simple, and suitable for rapid detection on site. The advantages and disadvantages of the above common methods are shown in Table 11.3.

Table 11.3 The advantages and disadvantages of common methods applied for detection of foodborne pathogens

Detection method	Advantage	Disadvantage
Traditional cultural, chemical and serologic detection	Accurate	Complex operation, time-consuming
Detection based on Resistance or conductance	Rapid	Only determination of the degree of pollution, inferior specificity
Direct bacterial counting	Rapid	Inferior specificity
Colorimetry based on dry weight	Rapid	Without specificity
Rapid cultivation based on paper	Accurate	Only applied for one microorganism
Colloidal gold immunochromatographic assay	Rapid, specific	Not sensitive and quantitative
Vitek automatic microorganisms analyzer system (VITEK-AMS)	Rapid, accurate, genus identification	Not suitable for identification of species
Vitek immune diagnostic assay system (VIDAS)	Wide detection range, rapid, high-throughput(30 samples)	Only used for preliminary screening
Gene chip (detection of target DNA)	Accurate, rapid, high-throughput	Expensive
Protein chip (detection of biotoxin)	Rapid, accurate, high-throughput	Expensive
UPT-POCT	Rapid, specific, accurate, quantitative	

11.4.2 Applications of UPT-POCT in Detection of Foodborne Pathogens

11.4.2.1 The Pre-treatment of Sample

The food and feces samples are required for enrichment. Take 0.6 g or 600 μl of food sample, or 0.3 g or 300 μl of stool sample into 5 ml of medium for enrichment, and then cultured for 5 h or overnight at 37 °C. 100 μl of supernatant of the culture is added to one tube with sample-treating buffer for detection.

The suspected liquid, powder, solid and organ samples are not required for enrichment. (1) For suspicious liquid sample, 100 μl are directly added to one tube with sample treating buffer for detection; (2) For suspicious powder sample, the wet cotton swab was first dipped in the tube with sample-treating buffer, and then used to obtained a proper amount of powder on the contaminated surface and washed in the same tube with sample-treating buffer for detection; (3) For suspicious solid and organ sample, an appropriate amount of sample (about 5 mm in diameter) is added into one tube with sample-treating buffer, and grinded into the homogenate for detection.

11.4.2.2 Parameters of UPT-POCT Reagents for Foodborne Pathogens

Taking the UPT-POCT detection kit for *Salmonella* spp. as an example, the parameters of UPT-POCT reagents for detection of foodborne pathogens are introduced. (1) Detection targets are common *Salmonella* spp. in samples, including *S. paratyphi* A, B and C, *S. enteritidis*, *S. typhi*, *S. typhimurium*, and *S. choleraesuis*. (2) It can be stored at 4–30 °C, valid for 18 months. (3) Applicable instruments are UPT biosensor and UPT-3A series analyzer. (4) For detection, 100 μl of the treated sample is added into the strip, and the results can be read after 15 min. (5) For qualitative detection, the result is judged as negative when the value of "results" is zero, and positive when the value is greater than 0. For quantitative detection, the meaning of the number of the "result" (Y) is that the concentration of target pathogens in sample added to the strip is 10^Y CFU/ml.

11.4.2.3 The Product of UPT-POCT for Detection of Foodborne Pathogens

At present, there are as many as 17 commercial UPT-POCT detection kits for common foodborne pathogens have been successfully developed (Table 11.4).

Table 11.4 UPT-POCT products for detection of foodborne pathogens

No.	Product name
1	UPT-POCT detection kit for *S. paratyphi* A
2	UPT-POCT detection kit for *S. paratyphi* B
3	UPT-POCT detection kit for *S. paratyphi* C
4	UPT-POCT detection kit for *S. enteritidis*
5	UPT-POCT detection kit for *S. typhi*
6	UPT-POCT detection kit for *S. choleraesuis*
7	UPT-POCT detection kit for common *Salmonella* spp.
8	UPT-POCT detection kit for *V. cholerae* O1
9	UPT-POCT detection kit for *V. cholerae* O139
10	UPT-POCT detection kit for *E. coli* O157:H7

(continued)

Table 11.4 (continued)

No.	Product name
11	UPT-POCT detection kit for *L. Monocytogenes*
12	UPT-POCT detection kit for *S. aureus*
13	UPT-POCT detection kit for *S. aureus* enterotoxin B
14	UPT-POCT detection kit for *V. parahaemolyticus*
15	UPT-POCT detection kit for *C. sakazakii*
16	UPT-POCT detection kit for *Shigella* spp.
17	UPT-POCT detection kit for *C. jejuni*

The rapid detection method of UPT-POCT for foodborne pathogens is an important supplement to Food Microbiology Test (GB4789-2016) of National Standard for Food Safety. UPT-POCT is particularly suitable for preliminary screening and detection on site in grassroots for food safety because of its fast and convenient characteristics. It has been widely used in provincial and municipal medical institutions. In addition, it is included in the reference list of equipment for food safety risk monitoring, for provincial and municipal disease prevention and control institutions, according to the "Construction Plan for Food Safety Risk Monitoring Capability (Equipment Configuration)", which issued by the Ministry of Health and the National Development and Reform Commission in China.

References

Jagadeesan B, Schmid VB, Kupski B, McMahon W, Klijn A. Detection of Listeria spp. and L. monocytogenes in pooled test portion samples of processed dairy products. Int J Food Microbiol. 2019;289:30–39.

Lepuschitz S, Ruppitsch W, Pekard-Amenitsch S, Forsythe SJ, Cormican M, Mach RL, Piérard D, Allerberger F. Multicenter study of cronobacter sakazakii infections in humans, Europe, 2017. Emerg Infect Dis. 2019;25(3):515–22.

Liu C, Shi C, Li M, Wang M, Ma C, Wang Z. Rapid and simple detection of viable foodborne pathogen. Front Chem. 2019;7:124.

Pereira JG, Soares VM, Tadielo LE, dos Santos EA, Lopes GV, Pellegrini DD, Duval EH, da Silva WP. Foods introduced into Brazil through the border with Argentina and Uruguay: Pathogen detection and evaluation of hygienic-sanitary quality. Int J Food Microbiol. 2018;283:22–7.

Rajapaksha P, Elbourne A, Gangadoo S, Brown R, Cozzolino D, Chapman J. A review of methods for the detection of pathogenic microorganisms. The Analyst. 2019;144(2):396–411.

Vinueza-Burgos C, Baquero M, Medina J, De Zutter L. Occurrence, genotypes and antimicrobial susceptibility of Salmonella collected from the broiler production chain within an integrated poultry company. Int J Food Microbiol. 2019;299:1–7.

Xu H, Zhang W, Guo C, Xiong H, Chen X, Jiao X, Su J, Mao L, Zhao Z, Li Q. Prevalence, serotypes, and antimicrobial resistance profiles among salmonella isolated from food catering workers in Nantong, China. Foodborne Pathog Dis. 2019.

Yang W, Huang L, Shi C, Wang L, Yu R. UltraStrain: an NGS-based ultra sensitive strain typing method for. Front Genet. 2019;10:276.

Zhao Y, Wang H, Zhang P, Sun C, Wang X, Wang X, Yang R, Wang C, Zhou L. Rapid multiplex detection of 10 foodborne pathogens with an up-converting phosphor technology-based 10-channel lateral flow assay. Sci Rep. 2016;6:21342.

Zoellner C, Jennings R, Wiedmann M, Ivanek R. EnABLe: An agent-based model to understand Listeria dynamics in food processing facilities. Sci Rep. 2019;9(1):495.

Chapter 12
Application of UPT-POCT in Detection of Food Safety Related Mycotoxins

Yanzhao Li, Xingbo Ren, Hongrui Zhang and Yong Zhao

Abstract Mycotoxin contamination is an important cause of food and feed safety issues. Mycotoxin contaminated foods can cause acute or chronic poisoning, and even cause cancers in humans and animals. Technology that can rapidly and quantitatively detect mycotoxins is of great importance for ensuring food safety and economic development. This chapter presents mycotoxins and their hazards, and introduces UPT-POCT for detection of mycotoxins. Comparisons of UPT-POCT and other detection methods are discussed in terms of both the methodology and practical applications. The results demonstrate that UPT-POCT is a promising tool for rapid, sensitive, and quantitative detection of mycotoxins in the field.

Keywords UCNPs · Mycotoxins · Lateral flow assay · On-site detection

Mycotoxin is one of the most dangerous naturally occurring food contaminants; it is a secondary metabolite produced by the growth of filamentous fungi in food and feed. Currently, more than 400 types of mycotoxin have been found, and there may be more that need to be identified and understood (Milićević et al. 2010). Mycotoxin contamination commonly occurs in food and oil, such as peanuts, soya beans, maize, cereals, and dairy products. In addition to causing loss of nutrients and reduced product quality, mycotoxins can also cause acute or chronic poisoning of humans and animals, and even cause cancer (Pfohl-Leszkowicz 2009). According to estimates by the FAO, ~25% of the world's grain and crops are contaminated with mycotoxins, causing losses of hundreds of billions of dollars annually. Mycotoxin contamination is prevalent in China and poses a serious food safety risk. It not only introduces an

Y. Li · X. Ren · H. Zhang · Y. Zhao (✉)
Beijing Key Laboratory of POCT for Bioemergency and Clinic (No. BZ0329), Beijing 100071, People's Republic of China
e-mail: zhaoyong179@139.com

Y. Li · X. Ren · H. Zhang
Beijing Hotgen Biotechnology Co., Ltd., Beijing Z-Park Daxing Biomedical Industry Base, No. 9, Tianfujie, Beijing 102600, China

Y. Zhao
Beijing Institute of Microbiology and Epidemiology, Academy of Military Medical Sciences, No. 20, Dongdajie, Fengtai District, Beijing 100071, China

R. Yang (ed.), *Principles and Applications of Up-converting Phosphor Technology*,
https://doi.org/10.1007/978-981-32-9279-6_12

important threat to the health of the general population but also seriously affects national economic development.

12.1 Mycotoxins and Their Hazards

Owing to the serious harm caused by mycotoxins, countries around the world have formulated corresponding limit standards and regulations. The European Commission announced standards for aflatoxins in foods more than 10 years ago: the total amount of aflatoxin in edible peanuts and their products must be less than 4 μg/kg, of which B1 must be ≤2 μg/kg. In 2017, China released the latest "Food Safety National Standards for Limits of Mycotoxins in Foods" (GB2671-2017), which replaced the standard (GB-2761-2011) implemented in 2011. Mycotoxins are listed as the focus of food safety monitoring. Common mycotoxins include aflatoxin, zearalenone, deoxynivalenol, ochratoxin A, T2 toxin, and fumonisin, amongst others.

12.1.1 Aflatoxin

Aflatoxins (AFA) are a series of structurally similar compounds (B1, B2, M1, and M2, etc.) primarily produced by *Aspergillus flavus* and *Aspergillus parasiticus*. Aflatoxins are widely known to contaminate products such as corn, peanuts, milk, and their by-products, with peanuts being the most commonly contaminated product (Wu et al. 2016). The most toxic aflatoxin subtype is aflatoxin B1, which is a Class I carcinogen as designated by WHO. It is one of the three major causes of liver cancer (Smela et al. 2001), and is the focus of food and feed testing. Aflatoxin B1 contamination mainly affects peanuts and corn, which can be ingested by people and cause disease.

12.1.2 Zearalenone

Zearalenone (ZEN) is a mycotoxin produced as an estrogenic metabolite by Fusarium fungi. Livestock, particularly pigs, can produce excessive amounts of estrogen after eating Zearalenone-contaminated feed (Schoevers et al. 2012), causing severe deterioration in fertility. In addition, Zearalenone also has a carcinogenic effect. Studies have shown that Zearalenone can bind to human estrogen receptors, stimulate the growth of breast cancer cells, and cause human breast cancer (Belhassen et al. 2015; Parandin et al. 2015).

12.1.3 Deoxynivalenol

Deoxynivalenol (DON), also known as vomiting toxin, is mainly produced by *Fusarium graminearum* and *Fusarium oxysporum*. DON can disrupt normal cell function by inhibiting protein synthesis (Pestka and Smolinski 2005), and cause a series of poisoning symptoms, such as loss of appetite, vomiting, diarrhea, ataxia, and slowness of movement; in severe cases it even causes death. All animal species evaluated to date are susceptible to DON; with pigs being the most susceptible (Pestka 2007). DON is also toxic to embryos and has mutagenic effects (Malekinejad et al. 2007).

12.1.4 Ochratoxin A

Ochratoxin A (OTA) is a secondary metabolite produced by Aspergillus and Penicillium (Ringot et al. 2006). OTA contaminated feed can cause severe diseases in animals. Pigs and poultry are the most susceptible to OTA. It mainly damages the kidneys and liver, and has teratogenic, carcinogenic, mutagenic, and immunosuppressive effects.

12.2 Current Status of Mycotoxin Detection Technology

Currently, the detection methods for mycotoxins primarily include high performance liquid chromatography (HPLC) (Arranz et al. 2004), liquid chromatography-mass spectrometry (LC/MS) (Hickert et al. 2015), enzyme-linked immunosorbent assay (ELISA) (Rossi et al. 2012), and colloidal gold lateral flow assay (LFA) (Table 12.1) (Maragos et al. 2010). Among them, chromatographic and mass spectrometry methods are the current gold standard. However, these techniques are not suitable for rapid detection and large-scale screening owing to their complex operation, time-consuming sample pretreatment, and requirement for sophisticated instrumentation.

Table 12.1 Comparison of current mycotoxin detection methods

Characteristics	Colloidal gold LFA	ELISA	HPLC	LC/MS
Time	0.5 h	Hours	Hours	Hours
Sensitivity	Low	Moderate	High	High
Sample pretreatment	Easy	Easy	Complicated	Complicated
Quantification	No/semi-quantification	Yes	Yes	Yes
Operation	Easy	Complicated	Complicated	Complicated
Trained operator required	No	Yes	Yes	Yes

Compared with the methods described above, colloidal gold LFA is a fast, stable, accurate, and efficient method for detecting mycotoxins. Qualitative or semi-quantitative results can also be obtained with a strip reader (Song et al. 2014). In addition, colloidal gold LFA does not require complicated sample pre-treatment or mycotoxin extraction with 70% methanol or ethanol. For example, methanol and ethanol can be replaced with water to extract mycotoxins (aflatoxin, fumonisin, and zearalenone) using Water Extraction Technology (WET) (Charm Sciences, USA) (Salter et al. 2006). WET technology allows rapid detection of mycotoxins in raw materials in 5 min under field conditions.

Another rapid method, time-resolved fluorescence LFA, was developed to achieve accurate quantitative detection of mycotoxins (Song and Knotts 2008; Tang et al. 2017). Time-resolved fluorescence LFA uses fluorescent signals (with wavelengths in the range 600–650 nm) as the quantitative detection signals. The fluorescence signal has greater intensity and better stability compared with the visible light of colloidal gold, allowing more sensitive detections. However, the major limitation of this method is that the background interference of the fluorescent signal can be very strong, which can affect the stability and accuracy of the detection result.

There is an urgent need in the market for a product that is fast, quantitative, accurate, stable, convenient, and highly sensitive for complex sample detection. UPT-POCT has developed very quickly in recent years, which has enabled the use of upconversion nanoparticles (UCNPs) as reporters in LFA detections (Niedbala et al. 2001; Yan et al. 2006; Yong et al. 2016). UCNPs demonstrate unique up-converting behavior. Compared with fluorescence, the up-converting phosphor signal has little background interference, and is very stable to environmental interference. Therefore, more accurate quantitative and sensitive detection can be achieved using UCNPs as reporters. The UPT-POCT method has unique advantages for mycotoxin detection:

- Accurate quantitative detection.
- Easy operation.
- Fast results. Quantitative test results can be obtained within 15 min.
- Safety. Operators do not need to handle toxin standards.
- Robust sample tolerance. The UPT-POCT assay has strong sample tolerance for various food matrices and environmental samples, which enables rapid sample detection in the field.

12.3 Application of UPT-POCT in Mycotoxin Detection

A variety of UPT-POCT detection kits for mycotoxins (Hotgen, China) have been developed (Table 12.2).

The principles and application of the detection kit are described in detail below, using vomiting toxin as an example.

Table 12.2 UPT-POCT kits for mycotoxins

Item	Package (test/kit)	Quantification range (ppb)	Time (min)
Aflatoxin B1	40	3–100	15
Aflatoxin M1	40	0.2–5	15
Zearalenone	40	50–1000	15
Vomiting toxin	40	100–6000	15
Ochratoxin A	40	60–2000	15
Fumonisin toxin	40	200–10,000	15
T2 toxin	40	60–2000	15

12.3.1 Principles of UPT-POCT Kits

The UPT-POCT assay is based on competitive immunochromatography. The test line (T line) of the NC membrane is coated with vomiting toxin antigen, and the control zone (C line) is coated with goat anti-mouse antibodies. UCNPs are covalently bonded to the monoclonal antibodies against vomiting toxin and are fixed in the conjugation pad of the strip. After sampling, the liquid flows along the strip under capillary action. During this process, the vomiting toxin antigen in the sample first binds to the UCNP-labeled antibodies against vomiting toxin. Then, the unbound UCNP conjugates are captured by antibodies at the T line. The UCNPs emit visible light signals (~541 nm) under irradiation with light at their excitation wavelength (980 nm). The ratio of the T line signal to the C line signal (T/C) is inversely proportional to the concentration of vomiting toxin in the sample. After a 15-min reaction, quantitative results can be obtained with the UPT biosensor.

12.3.2 Applicable Sample Types

The UPT-POCT kit is suitable for rapid quantitative detection of vomiting toxin in cereals and feeds. The strip can be set for different sample types (Table 12.3).

Table 12.3 Applicable sample types for the detection of vomiting toxin

Parameters	Sample types
A	Corn, grains, bran, soybean, cotton aphid, peanut, rapeseed, sub-powder
B	Wheat flour
C	Others

12.3.3 Detection Procedure

(1) Unpack the kit and place the test strip on a flat surface.
(2) Add 100 μl of liquid sample to the strip.
(3) Obtain the result from the UPT biosensor after a 15-min reaction.

12.3.4 Application and Evaluation

As shown in Table 12.4, the UPT-POCT kits for vomiting toxin are accurate and reliable; the deviation between the UPT-POCT result and the HPLC result is less than 10%. In addition, in a field test in a feed mill, the results showed that the accuracy and stability of the UPT-POCT kits were better than those of commercial ELISA kits (Beacon, USA), see Table 12.5.

Table 12.4 Comparison of the UPT-POCT kit and HPLC analysis results

Sample	UPT-POCT kits (ppb)	HPLC (ppb)	Deviation (%)
A	127	119	7
B	240	220	9
C	175	177	−1
D	286	280	2
E	600	650	−8

Table 12.5 Comparison of UPT-POCT kits and ELISA kits for detection of vomiting toxin in feed samples

Sample	UPT-POCT kits (ppb)			ELISA kits (ppb)		
	Test 1	Test 2	CV (%)	Test 1	Test 2	CV (%)
A	146.08	159.64	6.27	150.00	200.00	20.20
B	188.61	187.64	0.36	190.00	150.00	16.64
C	171.15	164.67	2.73	180.00	150.00	12.86
D	186.39	187.63	0.47	170.00	190.00	7.86
E	172.99	170.04	1.22	180.00	180.00	0.00
F	186.39	180.70	2.19	180.00	290.00	33.10

12.4 Conclusion

Mycotoxin contamination remains a prominent problem worldwide. Strengthening the research on prevention and control technology for mycotoxins in food, in particular establishing more rapid and accurate detection methods, is of great importance for ensuring food safety and economic development. UPT-POCT is a highly sensitive and accurate quantitative detection technology, which has been widely used in pathogenic bacteria detection and clinical diagnosis. However, the application in mycotoxin detection is still in its primary stage, and there remains a series of limitations to practical application. Further developments, such as improving the simplicity of mycotoxin extraction and developing multi-target detection kits, are still needed for the UPT-POCT technology to meet various testing requirements in the detection of mycotoxins in the field.

References

Arranz I, Baeyens WR, Van der Weken G, De Saeger S, Van Peteghem C. Review: HPLC determination of fumonisin mycotoxins. Crit Rev Food Sci Nutr. 2004;44(3):195–203.

Belhassen H, Jiménez-Díaz I, Arrebola JP, Ghali R, Ghorbel H, Olea N, Hedili A. Zearalenone and its metabolites in urine and breast cancer risk: a case-control study in Tunisia. Chemosphere. 2015;128(18):1–6.

Hickert S, Gerding J, Ncube E, Flett B, Cramer B, Humpf HU. A new approach using micro HPLC-MS/MS for multi-mycotoxin analysis in maize samples. Mycotoxin Res. 2015;31(2):109–15.

Malekinejad H, Schoevers EJ, Daemen IJ, Zijlstra C, Colenbrander B, Fink-Gremmels J, Roelen BA. Exposure of oocytes to the Fusarium toxins zearalenone and deoxynivalenol causes aneuploidy and abnormal embryo development in pigs. Biol Reprod. 2007;77(5):840–7.

Maragos CM, Busman M, Berthiller F. Rapid and advanced tools for mycotoxin analysis: a review. Food Addit Contam. 2010;27(5):688–700.

Milićević DR, Skrinjar M, Baltić T. Real and perceived risks for mycotoxin contamination in foods and feeds: challenges for food safety control. Toxins. 2010;2(4):572.

Niedbala RS, Feindt H, Kardos K, Vail T, Burton J, Bielska B, Li S, Milunic D, Bourdelle P, Vallejo R. Detection of analytes by immunoassay using up-converting phosphor technology. Anal Biochem. 2001;293(1):22–30.

Parandin R, Behnam M, Sisakhtnezhad S, Mahdavi N. In vitro evaluation of effects of Zearalenone and α-Zearalenol on MCF-7 and MDA-MB-468 cell lines of human breast cancer. Razavi Int J Med. 2015;3(4).

Pestka JJ. Deoxynivalenol: toxicity, mechanisms and animal health risks. Anim Feed Sci Technol. 2007;137(3):283–98.

Pestka JJ, Smolinski AT. Deoxynivalenol: toxicology and potential effects on humans. J Toxicol Environ Health Part B. 2005;8(1):39–69.

Pfohl-Leszkowicz A. Mycotoxins: a cancer risk factor. J Afr Cancer. 2009;1(1):42–55.

Ringot D, Chango A, Schneider YJ, Larondelle Y. Toxicokinetics and toxicodynamics of ochratoxin A, an update. Chem Biol Interact. 2006;159(1):18–46.

Rossi CN, Takabayashi CR, Ono MA, Saito GH, Itano EN, Kawamura O, Hirooka EY, Ono EYS. Immunoassay based on monoclonal antibody for aflatoxin detection in poultry feed. Food Chem. 2012,132(4).2211–6.

Salter R, Douglas D, Tess M, Markovsky B, Saul SJ. Interlaboratory study of the Charm ROSA Safe Level Aflatoxin M1 Quantitative lateral flow test for raw bovine milk. J AOAC Int. 2006;89(5):1327–34.

Schoevers EJ, Santos RR, Colenbrander B, Finkgremmels J, Roelen BA. Transgenerational toxicity of Zearalenone in pigs. Reprod Toxicol. 2012;34(1):110–9.

Smela ME, Currier SS, Bailey EA, Essigmann JM. The chemistry and biology of aflatoxin B(1): from mutational spectrometry to carcinogenesis. Carcinogenesis. 2001;22(4):535–45.

Song X, Knotts M. Time-resolved luminescent lateral flow assay technology. Anal Chim Acta. 2008;626(2):186–92.

Song S, Liu N, Zhao Z, Njumbe EE, Wu S, Sun C, De SS, Wu A. Multiplex lateral flow immunoassay for mycotoxin determination. Anal Chem. 2014;86(10):4995–5001.

Tang X, Li P, Zhang Q, Zhang Z, Zhang W, Jiang J. Time-resolved fluorescence immunochromatographic assay developed using two idiotypic nanobodies for rapid, quantitative, and simultaneous detection of aflatoxin and zearalenone in maize and its products. Anal Chem. 2017.

Wu LX, Ding XX, Li PW, Du XH, Zhou HY, Bai YZ, Zhang LX. Aflatoxin contamination of peanuts at harvest in China from 2010 to 2013 and its relationship with climatic conditions. Food Control. 2016;60:117–23.

Yan Z, Zhou L, Zhao Y, Wang J, Huang L, Hu K, Liu H, Wang H, Guo Z, Song Y. Rapid quantitative detection of Yersinia pestis by lateral-flow immunoassay and up-converting phosphor technology-based biosensor. Sens Actuators B Chem. 2006;119(2):656–63.

Yong Z, Xiao L, Wang X, Sun C, Wang X, Zhang P, Qiu J, Yang R, Lei Z. Development and evaluation of an up-converting phosphor technology-based lateral flow assay for rapid and quantitative detection of aflatoxin B1 in crops. Talanta. 2016;161:297–303.

Chapter 13
Application of UPT-POCT in Public Health Emergencies

Yanzhao Li, Xingbo Ren, Miao Jia and Yong Zhao

Abstract Rapid and reliable detection of infectious agents on site is essential for timely initiation of medical treatment and post-exposure prophylactic measures when public health emergencies occur. However, the referee standard for confirmation of infectious agents remains laboratory diagnosis, which is time-consuming and not available in the field. UPT-POCT technology is a versatile tool that requires limited resources and can realize rapid detection of infectious agents on site, providing timely information for the quick response to public health emergencies.

Keywords Public health emergency · POCT · UPT technology

13.1 Introduction

Since the 21st century, public health emergencies have occurred frequently in countries around the world, such as the outbreak of SARS in 2003, the H1N1 flu in 2009, the Ebola outbreak in 2014, and the outbreak of the African swine fever in 2018, etc. Public health emergencies have obvious unpredictability, seriously threatening people's health and social economic development. In order to minimize its impact, the risk factors should be quickly and accurately processed and eliminated after a public health emergency occurs, which puts higher requirements on the field detection technology.

When a public health emergency occurs, suspicious samples are usually transported to the laboratory for routine bioassay. However, it is difficult to provide

Y. Li · X. Ren · M. Jia · Y. Zhao (✉)
Beijing Key Laboratory of POCT for Bioemergency and Clinic (No. BZ0329), Beijing 100071, People's Republic of China
e-mail: zhaoyong179@139.com

Y. Li · X. Ren · M. Jia
Beijing Hotgen Biotechnology Co., Ltd., Beijing Z-Park Daxing Biomedical Industry Base, No. 9, Tianfujie, Beijing 102600, China

Y. Zhao
Beijing Institute of Microbiology and Epidemiology, Academy of Military Medical Sciences, No. 20, Dongdajie, Fengtai District, Beijing 100071, China

R. Yang (ed.), *Principles and Applications of Up-converting Phosphor Technology*,
https://doi.org/10.1007/978-981-32-9279-6_13

timely feedback because the transportation process and the testing process take a lot of time (Rongkard et al. 2016). Compared with conventional laboratory technology, POCT technology has the characteristics of fast detection, simple operation, and low requirements for instruments and operators, which is more suitable for application in public health events (Luppa et al. 2011). POCT technology can provide analytic results in the shortest time, providing important reference and evidence for relevant departments.

13.2 Diagnostic Methods for Public Health Emergencies

The gold standard for microbiological testing mainly relies on techniques such as separation and culture, morphological microscopy, and biochemical identification (Bloomfield et al. 2015). Such methods cannot meet the need for rapid and accurate detection of pathogens in the field because of its long period and complicated operation. In addition, there are some other conventional detection methods, such as enzyme-linked immunosorbent assay (ELISA) and polymerase chain reaction (PCR), which are faster than the gold standard method. However, these methods are also difficult to carry out in field conditions (Eriksson and Aspan 2007). The PCR method has been widely used in laboratories and hospitals. In order to detect the target gene of pathogens, it needs to extract and purify the genome in the sample first, and then detect the target gene with specific primers (Wisselink et al. 2017). The operation is complicated, and it takes 1–2 h to obtain the result. The colloidal gold based-immunoassay has the characteristics of low cost, simple operation, and rapid detection (within 15 min). It has been widely used as a rapid diagnostic tool, and is suitable for on-site detections. However, it is not sensitive and accurate enough, and it cannot provide quantitative result (Song et al. 2016).

UPT-POCT technology is a new detection technology that combines UPT technology, immunochromatography technology and biosensor technology. UPT-POCT technology uses UCPs as detecting labels, which has a unique up-conversion luminescence of emitting visible light (high energy) under the excitation of infrared light (low energy). This unique property makes it an excellent optic label with many advantages, such as no background interference, no quenching, and suitable for quantitative analysis. The development of UPT-POCT technology has obviously increased the sensitivity and stability of immunochromatography technology, and could complete quantitative detection with portable devices within 15 min (Hua et al. 2015).

13.3 The Application of UPT-POCT in Public Health Emergencies

UPT-POCT technology is suitable for the detection and analysis of various types of samples, including strong acid, alkali, high viscosity samples, powder samples, animal organ samples, and body fluid samples (Zhang et al. 2014; Li et al. 2009), which can basically meet the needs of on-site detection of public health emergencies. According to the report (Zhang et al. 2015), the UPT-POCT technology can also detect corrupt liver samples, which can meet the requirements for rapid detection of natural disease surveillance sites. The UPT-POCT kits for the detection of pathogens and viruses related public health emergency mainly includes *Yersinia pestis*, *Bacillus anthraci*, *Brucella sp*, *Burkholderia pseudomallei*, *Francisella tularensis*, ricin, abrin, Influenza A, Influenza B, Zika and Ebola virus.

The main parameters of the above UPT-POCT diagnostic kits are as follows: (1) the sample types can be soil, powder, animal organs and body fluid samples; (2) the instrument applied is the UPT-3A biosensor series, and the reaction time is 15 min; (3) the detection results can be printed and transferred to a PC; (4) the detection sensitivity of infectious bacteria can reach 2.0×10^3 CFU/ml; there is no cross between different pathogens; the detection has good repeatability; (5) the effective period of the kits is 18 months, and the storage conditions are 4 –30 °C, which can better meet the needs of on-site testing.

In summary, UPT-POCT technology has a wide range of applications in public health emergencies, providing rapid and timely detection results for related departments of disease control. At present, public health emergencies occur frequently in worldwide. The further improvement and promotion of UPT-POCT technology will provide more powerful technical support for the effective prevention and control of public health emergencies.

References

Bloomfield MG, Balm MN, Blackmore TK. Molecular testing for viral and bacterial enteric pathogens: gold standard for viruses, but don't let culture go just yet? Pathology. 2015;47(3):227–33.

Eriksson E, Aspan A. Comparison of culture, ELISA and PCR techniques for salmonella detection in faecal samples for cattle, pig and poultry. BMC Vet Res. 2007;3:21.

Hua F, Zhang P, Zhang F, Zhao Y, Li C, Sun C, Wang X, Yang R, Wang C, Yu A, et al. Development and evaluation of an up-converting phosphor technology-based lateral flow assay for rapid detection of Francisella tularensis. Sci Rep. 2015;5:17178.

Li L, Zhou L, Yu Y, Zhu Z, Lin C, Lu C, Yang R. Development of up-converting phosphor technology-based lateral-flow assay for rapidly quantitative detection of hepatitis B surface antibody. Diagn Microbiol Infect Dis. 2009;63(2):165–72.

Luppa PB, Müller C, Schlichtiger A, Schlebusch H. Point-of-care testing (POCT): current techniques and future perspectives. TrAC Trends Anal Chem. 2011;30(6):887–98.

Rongkard P, Hantrakun V, Dittrich S, Srilohasin P, Amornchai P, Langla S, Lim C, Day NP, AuCoin D, Wuthiekanun V, et al. Utility of a lateral flow immunoassay (LFI) to detect Burkholderia pseudomallei in soil samples. PLoS Negl Trop Dis. 2016;10(12):e0005204.

Song C, Liu C, Wu S, Li H, Guo H, Yang B, Qiu S, Li J, Liu L, Zeng H, et al. Development of a lateral flow colloidal gold immunoassay strip for the simultaneous detection of Shigella boydii and Escherichia coli O157:H7 in bread, milk and jelly samples. Food Control. 2016;59:345–51.

Wisselink HJ, Cornelissen J, van der Wal FJ, Kooi EA, Koene MG, Bossers A, Smid B, de Bree FM, Antonis AFG. Evaluation of a multiplex real-time PCR for detection of four bacterial agents commonly associated with bovine respiratory disease in bronchoalveolar lavage fluid. BMC Vet Res. 2017;13(1):221.

Zhang P, Liu X, Wang C, Zhao Y, Hua F, Li C, Yang R, Zhou L. Evaluation of up-converting phosphor technology-based lateral flow strips for rapid detection of Bacillus anthracis Spore, Brucella spp., and Yersinia pestis. PloS one. 2014;9(8):e105305.

Zhang P, Hua F, Yu X, Qu F, Xie H, Zhao Y, Zhao X, Jin L, Yu A, Cui B, Zhou L. Rapid detection of Yersinia pestis antigen from decomposed rodent viscera using an up-converting phosphor technology-based lateral-flow assy. Infect Dis Transl Med. 2015;1(2):58–60.

Chapter 14
Application of UPT-POCT in Import and Export Quarantine

Yanzhao Li, Honggang Zhang and Pingping Zhang

Abstract The objectives of port inspection and quarantine are to screen a large number of people for many types of diseases within a short window of time, deal with infectious diseases, respond to public health emergencies and bioterrorism attacks, and safeguard major international activities. UPT diagnostic reagents are accurate, low cost, easy to operate, easily mastered by short-term training, and allow instant detection without complex sample treatments. The UPT detection kits for antibodies against Zika virus, *Y. pestis*, antibody against *Y. pestis*, and Ebola virus have been used by the entry–exit inspection and quarantine bureaus for training and application in many areas in China.

Keywords Port · Inspection · Quarantine · UPT detection kit · Infectious diseases

Import and export inspection and quarantine institutions are part of the national defense in the field of public health. They strictly control the spread of high-risk infectious diseases among the border ports to ensure the health of personnel and sustain normal international trade. The entry–exit Health Quarantine Departments at each administrative level focus on carrying out proactive and preventive quarantine inspections; conducting inspections of entry–exit personnel and international navigation vehicles, baggage, cargo, and postal parcels; and conducting disease surveillance and health supervision at border ports (Wang 2004). Suspected infections from epidemic areas must be isolated and observed, and bags, containers, vehicles, or

Y. Li · H. Zhang · P. Zhang (✉)
Beijing Key Laboratory of POCT for Bioemergency and Clinic (No. BZ0329), Beijing 100071, China
e-mail: acedreams@126.com

Y. Li · H. Zhang
Beijing Hotgen Biotechnology Co., Ltd., Beijing Z-Park Daxing Biomedical Industry Base, No. 9, Tianfujie, Beijing 102600, China

P. Zhang
Beijing Institute of Microbiology and Epidemiology, Academy of Military Medical Sciences, No. 20, Dongdajie, Fengtai District, Beijing 100071, China

R. Yang (ed.), *Principles and Applications of Up-converting Phosphor Technology*,
https://doi.org/10.1007/978-981-32-9279-6_14

articles suspected of being contaminated must be treated to prevent the possible spread of contamination (Switzerland 2005). In China, the law enforcement administrative department for quarantine, certification, and standardization of entry–exit commodities, health, animals and plants, and import and export of food, is the General Administration of Quality Supervision, Inspection and Quarantine (AQSIQ).

14.1 Detection in Port Quarantine

In the 14th century, world pandemics of infectious diseases, such as plague and cholera, and the growth of marine navigation, led to the implementation of the first international health quarantine practice (Yang 1995). Italy was the first country to stipulate that only foreign ships that had been berthed for 40 days at sea and had no morbidity among crew members could enter its ports, and it demanded disinfection of products that any patients had come into contact with. This type of sanitation inspection measure plays a major role in controlling epidemics. The term "quarantine" is derived from the Latin word "quarantum", which means "40 days" (Li 2000). By the beginning of the 20th century, health quarantines for land routes had also been developed based on those that applied at ports, and International Health Regulations (revised in 2005) had replaced regulations created by individual countries or by mutual agreements (Wang et al. 2011). China's port health quarantine originated from quarantine measures implemented by Shanghai and Xiamen Customs during the 1873 cholera pandemic (Yang 1995). By the time of the Second Sino-Japanese War in 1937, China's frontier health quarantine attained a world-class standard, requiring quarantine for cholera, smallpox, yellow fever, and typhus (Wang 2009a). The Frontier Health and Quarantine Regulations of the People's Republic of China were issued in 1957, making a substantial contribution to the eradication of smallpox, as well as prevention of plague and cholera. Since the Chinese Economic Reform, port health quarantine has the integrated functions of quarantine inspection, disease detection, and health supervision (Wang 2009b).

The main objective of port quarantine is to screen a large number of people for many types of diseases within a short window of time. In addition to routine inspection and administrative enforcement for personnel, animals, plants, and food, port quarantine must also monitor infectious diseases and respond to public health emergencies and terrorist attacks. (1) Conventional quarantine is essential. For example, in the Middle Ages, with no conventional quarantine, many epidemics broke out on the European continent over a long period of time, resulting in a large number of deaths (Gao 2003). Modern routine medical inspections for entry–exit passengers can impede the transmission of diseases to the greatest possible extent (He et al. 2004). (2) The port quarantine office is at the forefront of preventing invasive harmful factors; however, infectious diseases in the latent period often have no symptoms. Therefore, the enforcement activities of the health and quarantine departments are primarily based on detection technology (Lei et al. 2007). (3) Global public health issues are involved. Infectious diseases that are now controlled in developed countries are still prevalent in developing countries and become potential sources for global infection. Large-scale tourist flows lead to the movement of microorganisms, and microorganisms that are not pathogenic to a particular race may cause diseases in other races. In addition, international trade and bird migration have further exacerbated the global threat of infectious diseases (Liu and Kuang 2007).

14.2 Demand for Development in the Field of Port Inspection and Quarantine

14.2.1 Three Functions of Port Inspection and Quarantine

The primary objective of port inspection and quarantine is to deal with infectious diseases. According to the World Health Organization, 32 new infectious diseases have been identified since the 1980s, including Acquired Immunodeficiency Syndrome (AIDS), cholera caused by *V. cholera* O139, and enterohemorrhagic diseases caused by *E. coli* O157:H7. Controlled infectious diseases have also been reemerging, such as plague, cholera, malaria, and tuberculosis (Zhou and He 2003). In addition, some rare infectious diseases have spread, such as Ebola hemorrhagic fever, Lhasa fever, and human Creutzfeldt-Jakob disease.

The second major task of port inspection and quarantine is to respond to public health emergencies and bioterrorism attacks (Xu and Huang 2007). Public health emergencies at border port inspections and quarantine include outbreaks of serious epidemic diseases, mass diseases of unknown origin, and grave poisoning incidents (Kong and Zhang 2008). Bioterrorism attacks (such as the mail attacks involving *B. anthracis* spores in the United States in 2001) can be very harmful, and they are often concealed and sudden, causing panic and resulting in lasting psychological effects. Port inspection and quarantine is an important line of defense for effectively intercepting such attacks.

The other major role of port inspection and quarantine is to safeguard major international activities. It is necessary to strengthen port inspection and quarantine during large-scale international events, such as the Olympic Games, World Expo, and Asian Games.

14.2.2 Port Inspection and Quarantine Demand for Detection Technology

The large quantities of items in port inspections and quarantine institutions must be screened within a short time frame; however, the detection equipment and resources are often basic and the staff lack high levels of specialized knowledge (Lei et al. 2007). These circumstances necessitate the application of detection methods, equipments, and reagents that are accurate, low cost, easy to operate, easily mastered by short-term training, and allow instant detection of results. In addition, the equipment should be small, and allow direct detection without complex sample treatments.

14.3 Up-Converting Phosphor Technology

14.3.1 UPT Test Kits Developed for Port Inspection and Quarantine

UPT test kits have been developed for port inspection and quarantine for detection of bacterial and viral infectious diseases, such as plague and Zika. Using up-converting phosphor nanoparticles (UCNPs) as bio-tracers for immunochromatography, the optical signal derived from the immune reaction can be collected and analyzed by up-converting phosphor technology (UPT), leading to accurate quantitative detection results. The components of a test kit include a test strip, bottles of sample treatment buffer, plastic droppers, cotton swabs, and instructions. The kit can be stored at 4–30 °C, and the period of validity is 18 months. The detection processes are as follows: (1) Unpack the test strip from the aluminum foil packaging and place on a flat surface. (2) Write the number of the sample on the shell of the test strip. (3) Mix the sample with sample-treatment buffer. (4) Add 100 μl of the treated sample to the well and wait for 15 min. (5) Place the strip in the UPT biosensor to obtain the results.

14.3.2 UPT Test Kit for Diagnosis of Plague

Plague is a serious infectious disease caused by *Y. pestis*, which is characterized by acute onset, high infectivity, and high mortality. It is a Class A infectious disease, as designated by the Law of the People's Republic of China on the Prevention and Treatment of Infectious Diseases. The three world pandemics of plague in history were profound disasters for mankind. Plague is a natural epidemic disease, with 17 provinces of natural foci in China. In 2000, plague was recognized by the World Health Organization as a re-emerging disease, and the International Health Regulations (2005) declared the disease as an international quarantine infectious disease (Cong and Zhang 2009; Li et al. 2009). Currently, the UPT detection kits for *Y. pestis* and antibodies against *Y. pestis* are used for diagnosis of plague based on the double antibody or antigen sandwich method.

14.3.3 UPT Test Kit for Diagnosis of Zika

Zika virus disease is a self-limiting acute infectious disease caused by the Zika virus (Gubler and Markoff 2007), which is mainly transmitted by Aedes mosquitoes. The main clinical symptoms are fever, rash, joint pain, and conjunctivitis, which rarely cause death; but Zika virus has attracted international attention owing to its relation to neonatal microcephaly and other neurological diseases (Schulerfaccini 2016). Up to February 2016, Zika virus had emerged in more than 40 countries and regions, with Brazil being the most affected country (Dick et al. 1952; Hayes 2009; Duffy et al. 2009; Ioos et al. 2014; Petersen et al. 2016; Tappe et al. 2013). China's Health Planning Commission has formulated the "Zika virus disease prevention and control program" and "Zika virus disease diagnosis and treatment program". Cases of imported Zika virus infection have been detected in China. AQSIQ has put a strong emphasis on the prevention and control of Zika virus disease in ports, which has strengthened the epidemic prevention effectively preventing the import of Zika virus disease.

The UPT detection kit for IgM antibodies against Zika virus is used for the qualitative detection of Zika virus IgM antibody in serum and plasma samples. The test line (T) of the nitrocellulose (NC) membrane of the strip is coated with recombinant antigen of Zika virus, and the control line (C) is coated with goat anti-rabbit IgG. During chromatography, the detected antibody in the sample is first bound by the UCNPs-rabbit anti-human IgM antibody complex, which is then captured by the T-band to form a recombinant antigen of Zika virus-IgM antibody-rabbit anti-human IgM antibody-UCNPs complex. A goat anti-rabbit IgG-rabbit anti-human IgM antibody-UCNPs complex is formed in the C-band.

14.3.4 UPT Test Kit for Diagnosis of Influenza

Influenza caused by H5 subtype influenza virus is primarily transmitted through the respiratory tract, and can also be spread through intimate contact with poultry and its secretions and excretions, as well as through water contaminated with viruses. H5N1 is a highly pathogenic avian influenza virus. H5N1 was first discovered in a human specimen in 1997. The Law of the People's Republic of China on the Prevention and Control of Infectious Diseases lists human infection highly pathogenic avian influenza as a Class B infectious disease, but specifies that the prevention and control measures appropriate for Class A infectious diseases should be applied if there is an outbreak.

The UPT detection kit for H5 subtype influenza virus is used for qualitative detection of influenza A virus in nasal or throat swab samples. This detection kit is based on the double antibody sandwich method. The test line (T) of the NC membrane of the strip is coated with antibody against H5 subtype influenza virus, and the control line (C) is coated with goat anti-mouse IgG. During chromatography, the antigen of interest is first bound by the UCNPs-antibody against H5 subtype influenza virus complex, which is then captured by the T-band to form an antibody-antigen-antibody-UCNPs complex. Goat anti-mouse-antibody against H5 subtype influenza virus-UCNPs complexes were formed in the C-band.

14.3.5 UPT Test Kit for Diagnosis of Ebola

Ebola hemorrhagic fever caused by Ebola virus is a serious infectious disease with a mortality rate of more than 50%, is highly contagious, and there is no relevant vaccine (Liu et al. 2012; Bente et al. 2009). The Ebola virus is classified as a class four high-risk virus by the World Health Organization, and the live virus must be handled in a level four biosafety laboratory (Xu and Qin 2010). According to the US Centers for Disease Control and Prevention, Ebola hemorrhagic fever is defined as a Class A disease, and Ebola virus is considered a potential biological warfare agent (Pourrut et al. 2005). The outbreak of Ebola hemorrhagic fever involved many parts of Africa (Pourrut et al. 2005; Colebunders and Borchert 2000; Peters and LeDuc 1999), and a large-scale outbreak occurred in West Africa in 2014.

A UPT detection kit for Ebola virus is used for qualitative detection of Ebola virus in serum and plasma samples. This detection kit is based on the double antibody sandwich method. The test line (T) of the NC membrane of the strip is coated with antibody against Ebola virus, and the control line (C) is coated with goat anti-mouse IgG. During chromatography, the antigen of interest is first bound by the UCNPs-antibody against Ebola virus complex, which is then captured by the T-band to form an antibody-antigen-antibody-UCNPs complex. Goat anti-mouse IgG-antibody against Ebola virus-UCNPs complexes are formed in the C-band.

14.4 Application

14.4.1 Current Detection Methods

Taking the diagnostic criteria for plague (ws 279–2008, Health Industry Standard of the People's Republic of China issued by the Ministry of Health) as an example, the current methods for detecting infectious diseases include clinical examination, isolation and identification of bacteria, polymerase chain reaction (PCR) detection, colloidal gold immunochromatography assay, enzyme-linked immunosorbent assay (ELISA), and reversed phase blood coagulation testing. (1) Clinical examination: the symptoms include high fever, rapid increase of white blood cells, chest pain, hemoptysis, and bloody diarrhea. However, the clinical symptoms of infectious diseases are similar to those of other causes. For precise diagnosis, clinical examination should be supported by an additional detection method. (2) Isolation and identification of

Table 14.1 List of the UPT detection kits used in entry–exit inspection and quarantine bureaus

No.	Department	Product name	Location
1	Wenshan entry–exit inspection and quarantine bureau	UPT detection kit for IgM antibodies against Zika virus UPT detection kit for *Y. pestis*	Tianbao port
2	Honghe entry–exit inspection and quarantine bureau	UPT detection kit for IgM antibodies against Zika virus UPT detection kit for *Y. pestis*	Jinshuihe port
3	Pu'er entry–exit inspection and quarantine bureau	UPT detection kit for Ebola virus	Meng'a port
4	Mengla entry–exit inspection and quarantine bureau	UPT detection kit for antibodies against *Y. pestis* UPT detection kit for *Y. pestis*	Mohan port
5	Xishuangbanna entry–exit inspection and quarantine bureau	UPT detection kit for IgM antibodies against Zika virus UPT detection kit for antibodies against *Y. pestis*	Daluo port
6	Ruili entry–exit inspection and quarantine bureau	UPT detection kit for antibodies against *Y. pestis*	Jiegao port
7	Ruili entry–exit inspection and quarantine bureau	UPT detection kit for antibodies against *Y. pestis*	Wanding port

bacteria: taking blood, pus, sputum, cerebrospinal fluid, lymph node puncture, and other materials for inspection, isolation of *Y. pestis* in combination with positive results for the *Y. pestis* phage lysis test provides bacteriological determination. In addition, injected test animal death and re-isolating the bacteria in the dead animal are used for determination of the highly virulent strains of *Y. pestis*. The isolation and culture of bacteria method is the gold standard for pathogen diagnostics, however it is time-consuming. (3) For PCR detection, the *fra* and *pla* gene fragments specific for *Y. pestis* are used as the target genes for amplification. PCR detection can give a diagnosis result within a few hours; however, the detection accuracy is highly dependent on the quality of the extracted target DNA. False positive results are often generated when PCR is applied in on-site testing. (4) The ELISA method can detect antibodies and antigens. Continuous washing steps ensure the specificity of the detection, but increase the risk of bacterial spread. (5) The results of the colloidal gold immunochromatography assay can be obtained in 15 min by simply loading the sample onto the strip, and the sensitivity is acceptable. (6) The reversed-phase hemagglutination test is an agglutination test of suspected serum using blood cells sensitized to antibody against the F1 antigen of *Y. pestis* F1, and the detection is rapid but the sensitivity is low.

14.4.2 The Application of UPT at Ports

The UPT detection kit is similar to the colloidal gold immunochromatographic assay. Its stronger tolerance for complex samples is a result of the enhanced bio-label performance of UCNPs compared with colloidal gold particles. The UPT detection kit for antibodies against Zika virus, *Y. pestis*, antibody against *Y. pestis*, and Ebola virus has been used by the entry–exit inspection and quarantine bureaus for training and application in many areas in China (Table 14.1).

References

Bente D, Gren J, Strong JE, Feldmann H. Disease modeling for Ebola and Marburg viruses. Dis Model Mech. 2009;2(1–2):12–17. https://doi.org/10.1242/dmm.000471 PubMed PMID: 19132113; PubMed Central PMCID: PMCPMC2615158.

Colebunders R, Borchert M. Ebola haemorrhagic fever—a review. J Infect. 2000;40(1):16–20. https://doi.org/10.1053/jinf.1999.0603 PubMed PMID: 10762106.

Cong X, Zhang C. Distribution of natural plague foci in the world and epidemics of human plague. Chin J Endemiol. 2009;28(4):357–360.

Dick GW, Kitchen SF, Haddow AJ. Zika virus. I. Isolations and serological specificity. Trans R Soc Trop Med Hyg. 1952;46(5):509–520. PubMed PMID: 12995440.

Duffy MR, Chen TH, Hancock WT, Powers AM, Kool JL, Lanciotti RS, et al. Zika virus outbreak on Yap Island, Federated States of Micronesia. N Engl J Med. 2009;360(24):2536–2543. https://doi.org/10.1056/NEJMoa0805715 PubMed PMID: 19516034.

Gao X. Epidemics and civilization. Xinhua Digest. 2003;9:61–65.

Gubler JD KG, Markoff L. Flaviviruses, in: Field Virology, Fifth Ed [M]. Philadelphia: Wolters Kluwer and Lippincott Williams & Wilkins. 2007.

Hayes EB. Zika virus outside Africa. Emerg Infect Dis. 2009;15(9):1347–1350. https://doi.org/10.3201/eid1509.090442 PubMed PMID: 19788800; PubMed Central PMCID: PMCPMC2819875.

He Z, Li Y, Wu H. Research on content of medical quarantine to entry–exit travelers. Port Health Control. 2004;9(6):30–32.

Ioos S, Mallet HP, Leparc Goffart I, Gauthier V, Cardoso T, Herida M. Current Zika virus epidemiology and recent epidemics. Med Mal Infect. 2014;44(7):302–307. https://doi.org/10.1016/j.medmal.2014.04.008 PubMed PMID: 25001879.

Kong J, Zhang L. Connotation analysis of emergency mechanism for public health emergencies. Soft Sci Health. 2008;22(2):166–169.

Lei H, Feng X, Liu Q. Discussion on viscounty and action of frontier health quarantine. Port Health Control. 2007;12(1):10–12.

Li C. Guide for China entry–exit inspection and quarantine. China procuratorial press. 2000:1193.

Li X, Wang M, Li S. The history, current situation, prevention and control of plague in China. Foreign Med Sci Sect Medgeography. 2009;30(03):125–128.

Liu W, Kuang W. The new traits of infectious diseases and how to deal with at frontier. Port Health Control. 2007;12(2):14–16.

Liu Y, Shi Z, Wang S, Wang Z, Ma Z. Ebola hemorrhagic fever. Chinese veterinary public health association;2012. p. 448–451.

Peters CJ, LeDuc JW. An introduction to Ebola: the virus and the disease. J Infect Dis. 1999;179 Suppl 1:ix–xvi. https://doi.org/10.1086/514322. PubMed PMID: 9988154.

Petersen E, Wilson ME, Touch S, McCloskey B, Mwaba P, Bates M, et al. Rapid spread of Zika virus in the Americas-implications for public health preparedness for mass gatherings at the 2016

Brazil Olympic Games. Int J Infect Dis. 2016;44:11–15. https://doi.org/10.1016/j.ijid.2016.02.001 PubMed PMID: 26854199.

Pourrut X, Kumulungui B, Wittmann T, Moussavou G, Delicat A, Yaba P, et al. The natural history of Ebola virus in Africa. Microbes Infect. 2005;7(7–8):1005–1014. https://doi.org/10.1016/j.micinf.2005.04.006 PubMed PMID: 16002313.

Schulerfaccini L. Possible association between Zika virus infection and microcephaly—Brazil, 2015. MMWR Morb Mortal Wkly Rep. 2016;65(3):59–62.

Switzerland. WHO Library Cataloging in Publication Data. World Health Organization International health regulations 2005. 2nd ed[Z]. 2008;9.

Tappe D, Rissland J, Gabriel M, Emmerich P, Gunther S, Held G, et al. First case of laboratory-confirmed Zika virus infection imported into Europe, November 2013. Euro Surveill. 2014;19(4). PubMed PMID: 24507467.

Wang X, Huang L, Hou T, Liu H. Research on the origin and evolution of the health quarantine. J Insp Quar. 2011;21(6):9–13.

Wang X. Research on histroty of frontier health quarantine of China (1). Port Health Control. 2009a;14(1):50–53.

Wang X. Research on histroty of frontier health quarantine of China (2). Port Health Control. 2009b;14(2):59–62.

Wang Y. Mode change of frontier heath quarantine and International Health Regulations. Chin J Frontier Health Quar Dec. 2004;27:52–53.

Xu J, Huang K. Implementing the International Health Regulations (2005) to strengthen supervision of port health quarantine. Sci Travel Med. 2007;13(4):49–50.

Xu L, Qin C. Progress on animal models of Ebola hemorrhagic fever. Chin J Comp Med. 2010;20(9):67–71.

Yang S. One hundred twenty years of China's health quarantine service. China J Med Hist. 1995;25(2):77–82.

Zhou J, He Y. Circumstance and countermeasures of Health Quarantine at Zhejiang Frontier Port after China's Entry into WTO. Chin J Front Health Quar Dec. 2003;26:89–92.

Chapter 15
Application of UPT-POCT in Detection of Drugs

Yanzhao Li, Yin Zhang and Pingping Zhang

Abstract Drugs are chemicals, anesthetics and stimulants with the addiction, including morphine, methamphetamine, tetrahydrocannabinol, ketamine, and methcathinone, etc. Due to the global expanding drug markets and drug driving problems, rapid screening detection has been performed for many countries. The detection windows of different test samples are different, such as urine, saliva, and hair. UPT-LF is of high safety, stability for preservation of evidence, accuracy with low false positive rate, portability, easily operating, and it has been widely used for detection of drugs from urine, saliva and hair samples in preliminary screening test for drug abuse. The commercial UPT-POCT reagents have been applied for detection of ketamine, methamphetamine, morphine, tetrahydrocannabinol, and carbamazepine, while the first three test reagents have obtained registration certification of the CFDA Class III medical device.

Keywords Drug abuse · Drug driving · Detection widow · Preliminary screening · UPT-POCT

Y. Li · Y. Zhang · P. Zhang (✉)
Beijing Key Laboratory of POCT for Bioemergency and Clinic (No. BZ0329), Beijing 100071, China
e-mail: acedreams@126.com

Y. Li · Y. Zhang
Beijing Hotgen Biotechnology Co., Ltd., Beijing Z-Park Daxing Biomedical Industry Base, No. 9, Tianfujie, Beijing 102600, China

P. Zhang
Beijing Institute of Microbiology and Epidemiology, Academy of Military Medical Sciences, No. 20, Dongdajie, Fengtai District, Beijing 100071, China

R. Yang (ed.), *Principles and Applications of Up-converting Phosphor Technology*,
https://doi.org/10.1007/978-981-32-9279-6_15

15.1 Characteristic of Drug Detection

15.1.1 Denomination of Drug

Drugs are chemicals, anesthetics and stimulants with the addiction, such as amphetamine-like drug (including methamphetamine, methylene dioxy-methamphetamine, etc.), opioid (including diamorphine, morphine, etc.), cannabis and ketamine. The characteristics of drugs are dependency, tolerance, high hazardness and illegal. Drug abuses of neotype of drugs by synthesis are much more than the traditional ones.

15.1.1.1 Morphine

Morphine (MOP) is an opioid receptor agonist, which accounts for about 4 to 21 percent of the opioid. Although morphine is a commonly used anesthetic in clinical practice because of its strong analgesic effect, it is easily addictive. Long-term use can lead to physical and psychological dependence, causing great harm to health. Heroin is an important derivative of morphine, namely morphine diacetate.

15.1.1.2 Methamphetamine

Methamphetamine (MET) is an extract or derivative of poppy gum with a molecular formula of $C_{10}H_{15}N$ and a molecular weight of 149. Its structural analogs are amphetamine and methylene dioxy-amphetamine (MDMA) etc. Methamphetamine, also known as ice and vigorous pills, has a pure white crystal appearance. When used as a drug, its forms are usually powder, liquid or pill. Methamphetamine has a strong stimulating effect on the central nervous system of the human body. Strong mental and physical dependences are generated after importation, reducing physical strength and immune function and seriously damaging hearts and brains, even leading to death. MDMA is a new type of amphetamine-like drug that has a stronger magical effect. 30 mg of methamphetamine can cause poisoning for the general population, but doses above 2000 mg can show excitatory effects for long-term abusers.

15.1.1.3 Tetrahydrocannabinol

Tetrahydrocannabinol (THC), also known as Δ9-tetrahydrocannabinol (Δ9-THC) or dronabinol, is a chemical synthetic drug. It is a major cannabis drug with strong hallucinogenic effect and psychology dependence, however body dependence is slight and tolerance is not easy to produce, therefore its harmfulness is relatively light compared with morphine and methamphetamine.

15.1.1.4 Ketamine

Ketamine (KET), commonly known as K powder, belongs to non-opioid anesthetic drugs. After importation, ketamine can cause separation of consciousness and sensation. It is characterized by stiffness, superficial sedation, forgetting and significant analgesia, and can show hallucinations, pleasures and violent tendencies.

15.1.1.5 Methcathinone

Methcathinone is a synthetic compound and usually imported in the form of a powder. Its addiction intensity is slightly weaker than that of methamphetamine. For first importation, 0.5 g of methcathinone can cause strong excitement and sleep loss. The abuse of methcathinone can cause many adverse consequences such as delusions, anxiety and abdominal pain, which cause serious damages to human health.

15.1.2 The Detection Character of Drugs

Drug abuse threatens human health, social stability and economic development around the world, becoming one of the most serious social problems. The World Drug Report in 2018 shows that the global drug market is expanding, and the number of deaths directly caused by drug abuse increased by 60% worldwide from 2000 to 2015. On June 25, 2018, "Report on China's Drug Situation in 2017" issued by the China National Narcotics Control Commission stated that China has solved 140,000 drug criminal cases and captured 892,000 tons of various types of drugs in 2017, therefore drug crimes had been effectively controlled, however the situation is still quite serious.

The circulation of drugs has seriously threaten people's living, and drug driving has become the next most dangerous activity besides driving after drinking (Verstraete et al. 2011; Xiang et al. 2016). An important part of the anti-drug work is the control of drug users, while detection of urine has been used as a routine mean for many years. Drug detection technology can effectively identify drugs and drug users and monitor the processes of drug rehabilitation, playing an important role in inhibiting drug abuse and epidemics.

The detection windows for different samples are different. Take methamphetamine as an example, methamphetamine in saliva mainly exists in the form of the original drug and has a high correlation with concentration of drugs in blood, which can be detected within half an hour to one day after importation. Although methamphetamine can be absorbed in the stomach and intestines, most of it is excreted in the form of the original drug in urine, and the half-life is 3–7 days. Because of a relatively low period for urine test, some drug addicts evade the inspection of community drug rehabilitation workstations by suspending importation four days before detection.

Therefore, in addition to urine and saliva, the detection of hair is also critical for the identification of drug addict (Shen et al. 2014).

15.2 Requirement for Drug Detection

Due to the great harm caused by drug driving, programs for rapid screening detection on the roadside have been established in many countries to effectively stop that behavior. At present, the composition and content must be provided for punishment measurement in the investigation and judge stage of drug-related crime cases, therefore, the rapid and quantitative detection of drugs is critical. The samples for drug testing are mainly urine, blood, saliva, hair and sweat etc. (Fabritius et al. 2013; Sarris et al. 2014), while tissues and organs in special cases. Sometimes, the sources of drugs should be traced. Sensitive, efficient, accurate and fast drug detection methods for screening are required.

15.3 UPT-POCT for Drug Detection

15.3.1 Principle of UPT-POCT for Drug Detection

Because drugs are small molecules, UPT-POCT detections for drugs are based on competition method. First, the UCP particles were coupled with monoclonal antibodies against drug and sheep IgG antibodies to prepare a conjugation pad of the strip. The detection band (T band) on NC membrane was coated with drug antigens, and the control band (C band) was coated with rabbit anti-goat antibodies. When the sample is dropped into sample well of the strip, the liquid is pulled up by the capillary force. The drug antigens in the sample are combined with the UCP-monoclonal antibody against drug complex during the chromatography, while the remaining UCP-monoclonal antibodies complex can be bind to the drug antigen coated on T band to from solid phase drug antigen-monoclonal antibodies-UCP complex, namely, the drug antigen in the sample competes with the drug antigen coated on the T band. Whether containing of drug antigens in sample or not, rabbit against goat-goat IgG-UCP complex is formed on C band.

The visible light signals can be generated by UCP particles under the excitation light of 980 nm, while the ratio of signals on T and C band (T/C) is used as a detection result. The concentrations of drugs in sample are inversely proportional to the T/C values. The strip can be analyzed by UPT biosensor, whether the sample contains drugs or not can be determined through comparison between the detection value and cutoff.

15.3.2 Technology Advance of UPT-POCT for Drug Detection

Colloidal gold strip reagent for urine detection is the earliest POCT reagent for rapid detection of drug, and it is widely used by the police. However, it bring confusion for the anti-drug work due to the high false positive rate, un-stability, incapability for quantitation, and subjective deviation caused by artificial color judgement, as well as difficulty for results preservation. UPT-POCT methods for rapid detection of drugs in urine overcome the above-mentioned shortcomings of colloidal gold detection reagents.

15.3.2.1 High Safety

The inorganic inert luminescent materials, as well as signal acquisition methods by infrared light excitation and visible light emission, make sure that UPT-POCT is not harmful to the relevant personnel and environment for detection on site.

15.3.2.2 The Uniqueness of Evidence

The results of drug testing, as physical evidence, need to be highly unique. The results of colloidal gold immunochromatographic assay are judged by naked eyes which can be easily interfered and counterfeited by humans. The luminescence phenomenon of UCP is a purely physical process generated by the internal structure, and its self-luminous performance is stable, avoiding the influence of various complex components in saliva, urine and hair on the detection of UPT-POCT in drug detection. A variety of samples can be directly tested after simple treatment, such as urine, saliva and hair. Therefore, the technical barriers for UPT-POCT play an important role for difficulty for fraud, and it can be used as a unique evidence.

15.3.2.3 High Accuracy and Specificity

The accuracy of colloidal gold detection method is easily influenced by the observer, the color of sample, pH value and the component of sample. For UPT-POCT, the sensitivity is high enough and up to the nanogram per milliliter, and specificity is excellent, due to the up-conversion luminescence of UCP particles with no background interference. Based on the result analysis and built-in standard curve by UPT biosensor, the detection is accurate.

15.3.2.4 Portability, Easily Operating, and Data Analysis

The UPT biosensor is compact and portable, with a weight of less than 3 kg. It is equipped with a touch screen and is easy to operate. The test result can be directly printed by the built-in printer or transmitted to the computer terminal for data analysis. The instrument can store up to 3000 test results.

The UPT biosensor can be supported by internal battery or external power supply, which is suitable for application on site. It can be used under the conditions of temperature 10–30 °C and humidity below 90%, and can be used at sea level of up to above 5000 m. In addition, UPT biosensor is equipped with a portable protective case for easy carrying.

15.3.3 Extensive Application of UPT-POCT for Drug Detection

For on-site rapid screening test, UPT-POCT can give qualitative results in 2 min for the samples of saliva, urine and hair, and further quantitative detection can be performed for suspected drug addict. For the analysis of corrupt biological samples, metabolites of drugs can be analyzed. For laboratories of hospital or provincial and municipal anti-drug department, the components and contents of drugs can be determined by UPT-POCT. It is also used for monitoring of drug addict in drug rehabilitation centers, as well as health examinations for military or special industry recruitment etc.

15.4 Application of UPT-POCT

15.4.1 Current Detection Methods for Drug Detection

The test methods are different for the types and the forms of drugs, as well as the detection objective. The extensively used methods are described as follows.

15.4.1.1 Routine Chemical Analysis Method

Suspicious drugs can be classified by routine chemical analysis method, usually by colorable reaction. Although simple and rapid, it has the disadvantages of low sensitivity, large detection error, and relatively subjective judgment for results analysis by the naked eye.

15.4.1.2 Thin-Layer Chromatography

This method is applicable to the grassroots public security institution because of simple operation, economical and practical. With thin-layer scanning, the types and components of drugs can be determined (Kuwayama et al. 2012).

15.4.1.3 Analysis Method Based on Large-Scale Instruments

The analysis method based on large-scale instruments is suitable for the analysis and identification of drug samples as evidence. Isolation and identification can be effectively performed using high-performance liquid chromatography or gas chromatography as separation methods, spectrum analysis (such as infrared spectroscopy, laser Raman spectroscopy and chemiluminescence analysis) or mass spectrometry as analysis methods, and data database as analysis means. The analysis method based on large-scale instrument has the advantages of high sensitivity, robustness for interference, and accurate quantification for each component of complex. However, due to the expensive instrument and high profession, it is mainly used as laboratory method, and not suitable for drug detection on sites (Rodrigues et al. 2008).

15.4.1.4 Capillary Electrophoresis Method

Capillary electrophoresis can provide basis data for the qualitative and quantitative determination of drugs. It included capillary zone electrophoresis and non-aqueous capillary electrophoresis, while the former can give quantitative results for heroin, methamphetamine and ecstasy, and the latter can analyze drugs with poor water solubility, such as alkaloids in opium. Capillary electrophoresis has the advantages of high efficiency, economy and requirement of trace sample volume, but separate selection and long-term automatic detection restrict its application (Isbell et al. 2015; Mikuma et al. 2015).

15.4.1.5 Immunological Analysis Method

Immunological analysis method is a preliminary screening method and must be confirmed by other methods, and it is generally used as on-site testing by grassroots public security department (Teerinen et al. 2014). Immunoassay method employs antibodies for detection of drugs, such as morphine, amphetamine, heroin and cannabis. With excellent specificity, immunological chromatography is especially suitable for rapid detection of drugs on site because it is simple, fast ($\leq$20 min), portable and easy to operate, such as colloidal gold immunochromatography and UPT-POCT method. Merely qualitative detection or semi-quantitative detection through color comparison is the disadvantage of colloidal gold immunochromatography. Quantitative results can be generated and directly printed by UPT-POCT (Hu et al. 2018) for detection on site, while suspected drug addict can be required to sign on the paper with printed results, making good service for law enforcement and greatly enhancing the confidence of law enforcement personnel.

15.4.2 Application of UPT-POCT for Drug Detection

15.4.2.1 UPT-POCT Reagents for Drug Detection

UPT-POCT is widely used in the preliminary screening test for drug abuse. UPT-POCT reagents for detection of ketamine, methamphetamine, morphine, tetrahydrocannabinol, and carbamazepine have been successfully developed, while the first three test reagents have obtained registration certification of class III medical device of China Food and Drug Administration (CFDA) (Table 15.1). There are mainly three sample types, including urine, hair, and saliva, and urine is the most widely used. At present, UPT-POCT projects for detection of drugs have been widely used in customs, border inspection, commodity inspection and other fields, especially in various law enforcement sites, such as marine anti-smuggling drugs activity. As the first rapid detection reagent for accurate quantitative detection of drugs in China, UPT-POCT plays a significant role in drug detection at primary level and first-line law enforcement.

The parameters of the reagents are described based on the three reagents obtained class III medical device certificate of the CFDA. (1) Expect usage is quantitative detection of drugs in saliva, urine or hair samples using competition method in primary screening of drug abuse. (2) Main components: UPT-POCT detection card, hair lysate solution or saliva collector. (3) Storage: Store at room temperature and the period of validity is 18 months. (4) Adaptation instrument: UPT biosensor (UPT-3A series); (5) Detection: 100 μL of the treated sample is applied and wait for 15 min at room temperature, the analysis time of UPT biosensor is less than 30 s; (6) Data analysis: Directly print the result report, or transmit the report to a computer terminal for data analysis; (7) Threshold and Specificity: The thresholds of UPT-POCT for various drugs in various samples were set mainly referring to the US Drug Abuse and Mental Health Services Administration and "Determination of amphetamine-

Table 15.1 UPT-POCT products for detection of drugs

No	Product	Product registration number[a]
1	UPT-POCT detection kit for ketamine	20173403282
2	UPT-POCT detection kit for methamphetamine	20173403284
3	UPT-POCT detection kit for morphine	20173403283
4	UPT-POCT detection kit for tetrahydrocannabinol	–
5	UPT-POCT detection kit for carbamazepine	–

[a]Approved by State Food and Drug Administration

Table 15.2 Threshold for UPT-POCT products

Detection target	UPT-POCT product	Concentration for positive suggestion (ng/mL)
Ketamine/methamphetamine	Reagent for urine	≥1000
	Reagent for saliva	≥50
	Reagent for hair	≥0.5
Morphine	Reagent for urine	≥300
	Reagent for saliva	≥15
	Reagent for hair	≥0.2

type stimulants, dolantin and ketamine in biological samples" in China Judicial Identification Technical Specification (SF/Z JD0107004-2016) (Table 15.2). UPT-POCT shows excellent specificity for naltrexone, buprenorphine pseudoephedrine, benzoylecgonine, diazepam, methadone, tramadol, gatifloxacin, codeine, aspirin, ethanol, glucose, and phenobarbitone at concentrations of 100 μg/mL, as well as 50 μg/mL of ranitidine, procaine and naloxone.

15.4.2.2 Pre-treatment of Different Sample for UPT-POCT

For pretreatment, saliva sample was collected using a saliva collector at no less than the position of the tick mark. 100 μL or 4 drops of saliva was drawing using a disposable pipette, and then add to the tube containing saliva sample treatment solution (0.01M PB), finally mixed by blowing for 10 times. Samples should be used as soon as possible after collected, it can be stored for 48 h at 2–8 °C if the test cannot be performed in time.

Urine sample can be directly used for detection if it is not turbid. Turbid urine sample should be centrifuged at 1000–2000 rpm for 3–5 min and the supernatant will be used for detection. Samples should be used as soon as possible after collection, while they can be stored for 48 h at 2–8 °C or at −20 °C for no more than 3 months with no more than 3 times of freeze-and-thaw.

For pretreatment of hair samples, the hairs should be cut into 1–2 mm segments using scissors; take an appropriate amount of shredded hair samples in a tube with hair extract solution (0.01 MPB). After ultrasonic treatment for five minutes, centrifugation at 1000–2000 rpm for 3–5 min will be performed, and the supernatant is taken for detection.

15.4.3 Application Prospect of UPT-POCT for Drug Detection

Currently drug detection reagents of UPT-POCT can comprehensively detect morphine, methamphetamine, ketamine and other drugs in saliva, urine and hair samples, and the detection types can be broadened at any time. The UPT-POCT test method can be performed less than 20 min with simple sample treatment, while the detection limit completely exceeds that required in the international standard. The drug detection system of "urine, saliva, and hair" has truely realized multi-dimensional and rapid screening on site, improving the accuracy of law enforcement on-site in anti-drug work. UPT-POCT will be useful in the fields of drug detection, forensic toxicology, clinical toxicology and stimulant detection.

References

Fabritius M, Chtioui H, Battistella G, Annoni JM, Dao K, Favrat B, Fornari E, Lauer E, Maeder P, Giroud C. Comparison of cannabinoid concentrations in oral fluid and whole blood between occasional and regular cannabis smokers prior to and after smoking a cannabis joint. Anal Bioanal Chem. 2013;405(30):9791–9803.

Hu Q, Wei Q, Zhang P, Li S, Xue L, Yang R, Wang C, Zhou L. An up-converting phosphor technology-based lateral flow assay for point-of-collection detection of morphine and methamphetamine in saliva. Anal. 2018;143(19):4646–4654.

Isbell TA, Strickland EC, Hitchcock J, McIntire G, Colyer CL. Capillary electrophoresis-mass spectrometry determination of morphine and its isobaric glucuronide metabolites. J Chromatogr B Analyt Technol Biomed Life Sci. 2015;980:65–71.

Kuwayama K, Tsujikawa K, Miyaguchi H, Kanamori T, Iwata YT, Inoue H. Rapid, simple, and highly sensitive analysis of drugs in biological samples using thin-layer chromatography coupled with matrix-assisted laser desorption/ionization mass spectrometry. Anal Bioanal Chem. 2012;402(3):1257–1267.

Mikuma T, Iwata YT, Miyaguchi H, Kuwayama K, Tsujikawa K, Kanamori T, Inoue H. The use of a sulfonated capillary on chiral capillary electrophoresis/mass spectrometry of amphetamine-type stimulants for methamphetamine impurity profiling. Forensic Sci Int. 2015;249:59–65.

Rodrigues WC, Wang G, Moore C, Agrawal A, Vincent MJ, Soares JR. Development and validation of ELISA and GC-MS procedures for the quantification of dextromethorphan and its main metabolite dextrorphan in urine and oral fluid. J Anal Toxicol. 2008;32(3):220–226.

Sarris G, Borg D, Liao S, Stripp R. Validation of an EMIT(R) screening method to detect 6-acetylmorphine in oral fluid. J Anal Toxicol. 2014;38(8):605–609.

Shen M, Chen H, Xiang P. Determination of opiates in human fingernail—comparison to hair. J Chromatogr B, Anal Technol Biomed Life Sci 2014;967(undefined):84–89.

Teerinen T, Lappalainen T, Erho T. A paper-based lateral flow assay for morphine. Anal Bioanal Chem. 2014;406(24):5955–5965.

Verstraete AKA, Jantos R, Skopp G, Gjerde H, Vindenes V et al. Driving under the influence of drugs, Alcohol and medicines (DRUID): per se limits—methods of defining cut-off values for zero tolerance. 2011, 2011. Accessed 20 July 2014.

Xiang X, Wang X, Jiang H, Chu C, Hao W. Drugs and driving in China: status and challenge. Int J Drug Policy. 2016;31:203–204.

Chapter 16
Application of UPT-POCT in Anti-bioterrorism and Biosecurity

Pingping Zhang

Abstract With the exception of toxins, bioterrorism agents are mainly microorganisms, many of which cause serious infectious diseases. Up-converting phosphor technology-based point-of-care testing (UPT-POCT) can detect bioterrorism agents from various samples with high sensitivity and specificity, in particular it shows robust performance for complicated samples, such as food, powder, viscera and grains. The tolerance of UPT-POCT to sample is based on the physical and luminescence stability of UCNPs, the stable covalent interaction between UCNPs and antibody, as well as the strong buffering capacity of the detection system. Reliable results can be obtained in a short time period using a portable biosensor by nonprofessionals owing to the simple nature of UPT-POCT operation and sample pre-treatment.

Keywords UPT-POCT · Bioterrorism agents · Operational safety · Performance evaluation

16.1 Introduction

Bioterrorism is an activity threatening public health with violence using biological methods for political or other purposes, while biological warfare is a military activity involving biological weapons. The pathogenic microorganisms or toxins in bioterrorism and biological warfare, termed biological warfare or bioterrorism agents (Porche 2002), can not only attack the susceptible human populations, but also susceptible animals and plants, causing significant economic losses. In addition, the widespread distribution of pathogenic microorganisms or toxins in natural reservoirs is a potential threat as a trigger of public health emergencies.

Significant attention has been focused on bioterrorism since the *B. anthracis* spore attack that followed the events of '9.11' in the USA (Gouvras 2002), while people are

P. Zhang (✉)
Beijing Key Laboratory of POCT for Bioemergency and Clinic (No. BZ0329), Beijing 100071, China
e-mail: acedreams@126.com

Beijing Institute of Microbiology and Epidemiology, No. 20, Dongdajie, Fengtai, Beijing 100071, People's Republic of China

R. Yang (ed.), *Principles and Applications of Up-converting Phosphor Technology*,
https://doi.org/10.1007/978-981-32-9279-6_16

also scared by frequent outbreaks of infectious disease in natural foci, such as plague and anthrax. Accurate detection and identification of the relevant microorganisms and toxins in routine surveillance or during public health emergencies, is the main line of defense for public security.

16.2 Overview of Bioterrorism Agents

16.2.1 Categories of Bioterrorism Agents

The great variety of bioterrorism agents were determined and defined by contracting parties to the Biological Weapons Convention in 1996, and slight changes have been made in subsequent meetings. Some bioterrorism agents, such as *Yersinia pestis*, *Bacillus anthracis*, *Brucella* spp., and *Coxiella burnetii*, are zoonotic pathogens that survive in reservoirs in natural foci and have become a potential long-term threat to public health. In addition, some toxins used as bioterrorism agents can be easily obtained from organisms in nature. With the development of molecular biology, new bioterrorism agents can be made by DNA recombination or cell fusion based on existing agents.

With the exception of toxins, bioterrorism agents are mainly microorganisms, such as bacteria, viruses, rickettsia, and chlamydiae. Some examples of bacterial bioterrorism agents include *Y. pestis* (Prentice and Rahalison 2007), *B. anthracis*, *F. tularensis*, *Brucella* spp., *B. pseudomallei*, *V. cholerae*, and *S. typhi*. *Y. pestis* and *V. cholerae* are pathogens for plague and cholera, respectively, which are two Class A infectious diseases as defined by the Law on Prevention and Treatment of Infectious Diseases issued by the People's Republic of China. Most viral bioterrorism agents are classified as RNA viruses, such as Marburg virus, Forest encephalitis virus, Hanta virus, Human immunodeficiency virus, SARS coronavirus, MERS coronavirus, Ebola virus, Spanish flu virus, H1N1 flu virus, Avian flu virus, Hepatitis C virus and Rabies virus; in contrast there are few DNA viruses, such as Hepatitis B virus. Rickettsial bioterrorism agents include *Coxiella burnetii*, *Rickettsia rickettsia*, and *Rickettsia prowazekii* among others, and they are strictly parasitic microorganisms. *Chlamydia psittaci* are the main chlamydial bioterrorism agents and can infect many bird species and humans. Biotoxin bioterrorism agents are noxious materials secreted by live organisms, such as plants, animals, microorganisms and insects, and are regarded as occupying the space between traditional biological and chemical terrorism agents. Common biotoxin bioterrorism agents, botulinum toxin and *S. aureus* enterotoxin, are secreted by microorganisms, while ricin and abrin are extracted from plants. There are two categories of biotoxins, one group are proteins or peptides such as ricin and abrin, and the other are small molecule toxins, such as afaltoxin and T-2 toxin (Wang 2011). Protein biotoxins combine an activity unit (unit A) and binding site (unit B). Unit A is the functional domain and unit B promotes the introduction of the toxin into cells.

Based on pathogenicity, bioterrorism agents can be categorized into lethal and incapacitating agents. Lethal bioterrorism agents have high mortality, for example, the mortality of septicemic plague and pulmonary tularemia can reach 90% and 60%, respectively, in the absence of antibiotic treatment. *Y. pestis*, *B. anthracis*, *F. tularensis* Type A, *B. pseudomallei*, Yellow fever virus, Smallpox virus, *Rickettsia rickettsii*, *Chlamydia psittaci*, and botulinum toxin are all lethal bioterrorism agents. Incapacitating bioterrorism agents can make people defenseless, examples include *Brucella spp.* and *Coxiella burnetii*.

16.2.2 Common Transmission Routes of Bioterrorism Agents

Bioterrorism agents can be transmitted by air, food, and water, and some can be transmitted person-to-person, such as *Y. pestis* (Begier et al. 2006), *B. anthracis*, and Ebola virus. The bioterrorism agents disseminated through air can be made into aerosols and threaten public health on a large-scale, examples include *Rickettsia rickettsii*, *Y. pestis* (Agar et al. 2009), and *B. anthracis* (Estill 2009). Some zoonotic pathogens can infect people through contact with infectious animals during slaughter and leather treatment, and animal husbandry, as well as contact between people (Begier et al. 2006). Transmission through food includes consumption of the meat or milk of infected animals, for instance, people can be infected by *V. cholerae* through contaminated seafood (Finelli et al. 1992). Water is also an important transmission medium and is the main transmission route for *V. cholerae* (Hill et al. 2011). In addition, some agents, for example, *B. anthracis* spore can survive in the silt at the bottom of a riverbed for decades, and *F. tularensis* can survive in the cold water of a river for months (Chitadze et al. 2009), seeking the chance for outbreak. Transmission media arthropods such as mosquitos, flies, lice, mites, and ticks, are widespread in nature. The transmission media of *Y. pestis*, *F. tularensis*, yellow fever virus, and *Rickettsia przewalskii* are flea, tick, mosquito and pediculus humanus corporis, respectively.

16.2.3 Perniciousness and Diagnosis of Bioterrorism Agents

The suspicious incidents caused by bioterrorism agents can easily lead to public panic because of their perniciousness. The misdiagnosis of the diseases caused by bioterrorism agents, rapid deteriorations for acute and serious infectious diseases, and the limitations of therapeutic means, all highlight the importance of preventing the release of bioterrorism agents.

Low pathogenic dose, diversity of pathogenic types, high mortality or disability rate, and the potential for widespread dissemination, are all characteristics of bioterrorism agents, as well as strong adaptability to the environment in the exposed zone resulting in long-term threat to public health. Several microorganisms can multiply rapidly in vivo, causing serious diseases, for instance, the infectious dose of *F. tularensis* type A is less than ten live cells. Bioterrorism agents attack the respiratory tract, digestive tract, skin, blood, and glands. Many of the agents cannot be handled by the immune system, and *Y. pestis* can even survive, proliferate and spread in macrophages (Lukaszewski et al. 2005). Acute and serious diseases occur after a short incubation period, often only several days. For instance, the symptoms of pneumonic plague include high fever, cold shivers, cough, chest pain, hemoptysis, and dyspnea, followed by serious poisoning symptoms and death in two to four days. The mortality of pneumonic and generalized plague is up to 30 ~ 60%. *Brucella* spp. with anti-phagocytic capsules can proliferate in the lymphatic system and then enter into the blood resulting in toxemia, and its ability to evade elimination by the body results in long-term joint pain, fatigue, and disability, dramatically decreasing the quality of life of the patient. The most catastrophic potential outcome is a worldwide epidemic. Three historic large-scale plague epidemics caused 160 million deaths (Prentice and Rahalison 2007), seven historic cholera epidemics involved 100 countries, and brucellosis is prevalent in 170 countries.

Diagnosis can be based on epidemiological history, clinical syndromes, etiology, and serology detection. Epidemiological history includes residency in the epidemic areas or entrance into these areas in the last two weeks before morbidity, bites by arthropods, and contact with or consumption of infected products or water. Clinical syndromes of some bioterrorism agents appear in isolation, however agents with many infection routes can cause various clinical syndromes. For example, the main clinical syndrome of cholera is diarrhea, however at least six syndromes have been found for tularaemia, including bubonic, pneumonic, gastrointestinal, and systemic (typhoidal and septicemic) tularaemia. The similarity of clinical syndromes between diseases caused by bioterrorism agents and by other factors, makes accurate pathogenesis diagnosis more difficult for doctors when the etiology is unknown. For example, patients with fever caused by *F. tularensis* are easily misdiagnosed with influenza (Simsek et al. 2012), while there are no differences the in clinical syndromes between pneumonia caused by *F. tularensis* or that arising from other causes (Stralin et al. 2002). Misdiagnosis and delayed treatment are responsible for lack of safeguard implementation and the subsequent dissemination of bioterrorism agents in medical institutions. Therefore, etiology detection is critical for determining and preventing infectious disease. However, the low numbers of pathogens in the initial stage of a disease are difficult to find by etiology detection, fortunately early diagnosis based on the detection of antibodies in blood is a plausible approach in practice.

16.2.4 The Therapy and Prevention of Bioterrorism Agents

Because of the great threat of bioterrorism agents to the lives of patients, essential therapy must be administered in time to avoid death and poor prognosis. For instance, water and electrolytes must be administered to patients infected by *V. cholerae* in time because severe dehydration and failure of microcirculation caused by diarrhea often occur during the progression of cholera. The prognosis is usually poor owing to the limited therapeutic means. Antibiotics are often excessively applied for saving the lives of patients infected by bacterial bioterrorism agents, such as streptomycin specific for plague and many antibiotics that are effective for tularaemia, however the side effects of this therapy method are osteoporosis and joint injury.

Eliminating infection sources, cutting-off transmission routes, and protecting susceptible patients, are all methods for prevention and control of the spread of bioterrorism agents. Quarantine is essential for infectious bioterrorism agents, including isolation of prime areas of disease outbreak, quarantine of patients and people who have gone to countries with epidemics, conflagration and deep interment of cadavers, use of disinfectant, and correct treatment of material from patients with high-temperature and high-pressure. Individual and environmental defenses should be enhanced when nursing and treating patients or infected animals, as well in the resulting treatment of cultures in scientific research. Pre-inoculation of vaccines is a good prevention method for protection of people in natural foci, doctors and scientific staff, for example, vaccines or attenuate strains of *Y. pestis*, *F. tularensis*, and *B. anthracis* can be inoculated by cutaneous scarification. However, no effective vaccine has been obtained for some bioterrorism agents, such as *Chlamydia psittaci*. There are some effective virus vaccines, such as Vaccinia vaccine (against Smallpox virus), Rabies vaccines, Hepatitis B vaccines, and Hantavirus vaccine. However, mutated viruses often emerge as RNA viruses, especially for retroviruses such as Human immunodeficiency virus, leading to some vaccine failure and the necessity for preparation of new vaccines against the mutated virus.

16.3 Characteristics of Detection in Bioterrorism and Biosecurity

Owing to the high pathogenicity of bioterrorism agents, as well as the high transmission capacity of microorganisms, essential isolation and protection measures must be carried out to ensure biosecurity during the process of detection. Prompt and accurate testing favors the detection of suspicious substances, so that an emergency signal can be issued or a false alarm can be revealed. In addition, multiplexed detections can be applied to unknown pathogens to improve efficiency and reduce environmental contamination caused by excessive operation.

16.3.1 Operational Safety

Protective measures during the detection process are essential to ensure individual and environmental security and prevent the spread of bioterrorism agents. Specialized laboratories for pathogenic microorganisms, or detection vehicles on site equipped with such laboratories are necessary for lethal bioterrorism agents. For detection on site, operators should wear rubber gloves, surgical masks, hats, and protective clothes. Isolation for patients and infected animals, and blockades of the contaminated region are essential. The screened suspicious matter should be sent to the laboratory for further identification in a biohazard marked container that is waterproof, breakage-proof, and high temperature and pressure resistant. According to the national standard for laboratory security, the defense levels of bio-laboratories are from 1 to 4. Experiments involving the most infectious bioterrorism agents must be performed by professionals using a biosafety cabinet in a level 3 or 4 laboratory. The measures for infection prevention in the laboratory include pre-inoculation with vaccine, wearing protective clothes, using biosafety cabinets for culture and infective material, high-pressure treatment for growth medium and contaminated gloves, as well as sprinkling disinfectant in the biosafety cabinet.

16.3.2 Detection Accuracy and Detection Time

The detection limit, specificity, and tolerance of detection methods determine their accuracy.

The detection limit for bioterrorism agents must be very low, ideally single cell, because of their high pathogenicity in low dose and the high capacity for proliferation of some microorganism bioterrorism agents. Pre-incubation of microorganism bioterrorism agents to increase detection rates is not permitted. Specificity is particularly critical for a detection method because there are incalculable microorganisms and other organisms in nature. No specific reaction should occur for substances or strains that share structural similarity, close genetic relatedness or similar transmission routes with the targets. A detection method must be available for different specimens within appropriate operational-error to ensure the stability. Various fresh or decomposed animal tissues obtained in natural foci and the white powders used by terrorists for concealment, such as flour, milk powder, and putty powder, are representative of the complicated specimens that the detection method will be confronted with. The errors brought by non-professional operation are also considered to ensure the detection accuracy in practice.

In addition to safety and accuracy, detection time should also be shortened. A short detection time is beneficial for therapy, cutting-off transmission routes, preventing the dissemination of contamination, and promptly eliminating negative social effects.

16.3.3 Multiplexed Detection

Multiplex detection has been developed to improve detection efficiency, and reduce sample volumes and operational handling. Compared with the detection of individual targets, multiplex detection reduces the workload and shortens detection times. In addition, multiplex detection can give results for many targets at once, therefore smaller sample sizes are sufficient, which is especially important for precious low-volume samples. Less operational handling minimizes the possibility of contamination of the inspector and environment with bioterrorism agents. Because of difficulties arising from multiple reactions in one system and simultaneous multiple signal extraction, multiplex detection still shows unsatisfactory performance in stability, anti-interference, and reproducibility. Despite these shortcomings, multiplex detection is urgently required for identification of unknown pathogens and simultaneous detection of multiple pathogens in one sample because multiple bioterrorism agents can be released in one sample, for example, *Coxiella burnetii*, *Chlamydia psittaci* and Influenza virus have been combined into aerosols and applied for biowarfare.

16.4 Requirements for Detection of Bioterrorism-Associated Agents

16.4.1 Field of Application

Bio-threats are becoming increasingly serious with developing technology because of reductions in cost, and improved transmission and operation. Many bioterrorism agents can be easily obtained from nature, such as zoonotic microorganisms from natural hosts, and ricin and abrin, which are easily prepared from plants. Modern fermentation technology promotes mass-production of microorganism bioterrorism agents (Yang 2008), while many toxins can be synthetized using chemical methods. Water, air conditioning systems, food, and letters have been used as the vectors for bioterrorism agents in bio-attack incidents, and the agents can appear in the form of powders and aerosols, amongst others.

The detection of bioterrorism agents includes surveillance in natural foci and treatment for public health emergencies. Surveillance, especially for common infectious disease, is an efficient measure for preventing disease outbreaks. The objects of the monitoring are different depending on the sort of disease. For example, rodents and fleas are natural reservoirs for *Y. pestis* (Prentice and Rahalison 2007); water and plankton are the natural environment and reservoir of *V. cholerea* respectively (Huq et al. 1995); and *B. pseudomallei* can survive in tropical and subtropical natural foci (Draper et al. 2010). Public health emergencies, such as bioterrorism attacks, laboratory releases, collective food poisoning, and concentrated outbreaks of cases, require prompt responses based on the detection results.

16.4.2 *Point of Care Testing*

According to the definition specified by the Committee of POCT, Chinese Association of Medical Equipment, point of care testing (POCT) is a detection method implemented on site, and it reports results in a short time period using portable analytical instruments and associated reagents. To satisfy the strict requirements of POCT, a detection method must be sufficiently rapid, sensitive, and specific.

Screening detection on site mainly relies on POCT methods. The short response time for surveillance in natural foci and during public health emergencies has provided powerful support for the prevention of disease outbreaks. The minimal operation procedures that are a feature of POCT methods make POCT applicable for detecting bioterrorism agents to reduce the release of the agents. Immunochromatography is well known as a POCT method for onsite screening of bioterrorism agents because it can give results from simple sample loading and the used strip can be directly disposed of after treatment with high temperature and pressure. The sensitivity and specificity of traditional immunochromatography methods are often too low because the physical interaction between antibodies and gold particles is fragile, and the results are analyzed by the naked eye. Up-converting phosphor technology-based point-of-care testing (UPT-POCT) as described below are based on the covalent conjugation of upconversion nanoparticles (UCNPs) and antibodies, and then the emission signal at 540 nm resulting from excitation at 900 nm can be transmitted into readable electrical signals by biosensors. Therefore, the UPT-LF assay can detect various samples with high sensitivity and specificity, in particular it shows robust performance for complicated samples.

16.5 UPT-POCT Assay Applied for Detection of Bioterrorism Agent

16.5.1 *Detection Mode*

The detection mode of UPT-POCT for bioterrorism agents hinges on the molecular size of the detection target. Double-antibody sandwich mode is used for detection of macromolecules, while competition mode is used for micro-molecules. In addition, the principle of the double-antigen sandwich mode is in parallel with that of the double-antibody sandwich mode for utilization of macromolecules, mainly antibodies.

UPT-POCT based on the double-antibody sandwich mode is mainly applied for the detection of bacterial antigens or protein toxins, such as *Y. pestis* (Yan et al. 2006), *Brucella* spp. (Qu et al. 2009), *B. anthracis* spore (Li et al. 2006), *F. tularensis* (Hua et al. 2015b), *B. pseudomallei* (Hua et al. 2015a), *V. cholerae* (Hao et al. 2017), *E. coli* O157 (Wang et al. 2007), ricin (Wang et al. 2016), and abrin (Liu et al. 2016). While UPT-POCT based on the double-antigen sandwich mode is applied for the detection

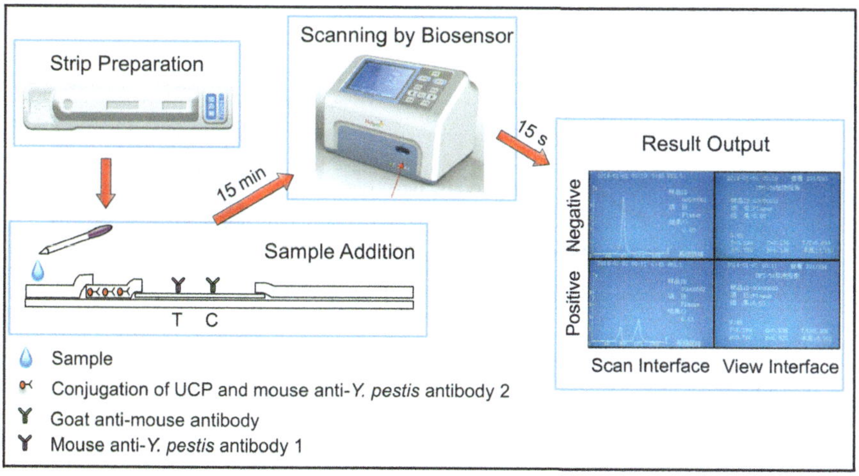

Fig. 16.1 Schematic for detection of *Y. pestis* by UPT-POCT based on the double-antibody sandwich mode

of antibodies against bioterrorism agents, such as antibodies against *Y. pestis* (Hong et al. 2010) and hepatitis B surface antibody (Li et al. 2009). For example, in the detection of *Y. pestis* (Fig. 16.1); mouse anti-*Y. pestis* antibody 1 is immobilized on the nitrocellulose membrane as the test (T) line, while mouse anti-*Y. pestis* antibody 2 is combined with UCNPs covalently. For positive detection, *Y. pestis* in samples is captured by UCNPs-mouse antibody 2 conjugates, and then flows forward and is captured by mouse antibody 1 at the T line of the nitrocellulose membrane, forming T line-mouse antibody 1-*Y. pestis*-mouse antibody 2-UCNPs conjugates. UCNPs can emit visible light signals after excitation by infrared light, and the intensity of the signals is in direct proportion to the concentration of *Y. pestis*. Whether there are *Y. pestis* present or not, goat anti-mouse antibody-mouse antibody 2-UCNPs conjugate will be formed on the control (C) line on the nitrocellulose membrane for quality control. The signal ratio between the T and C lines, the T/C ratio, is defined as the detection result, and T/C values increase with the increase in quantity of *Y. pestis* in the sample. In the double-antigen sandwich mode, bioterrorism antigens, such as F1 antigen of *Y. pestis*, are dispensed in the T line and combined with UCNPs respectively, and then the corresponding antibody is detected.

Competitive mode is used to detect micro-molecular matter that is too small to be detected by double-antibody sandwich mode. UPT-POCT based on competitive mode is applied for the detection of mycotoxin, such as aflatoxin B1 (AFB1) (Zhao et al. 2016), aflatoxin M1 (Liu et al. 2014) and T-2 toxin. Using the detection of aflatoxin B1 (AFB1) as an example (Fig. 16.2), AFB1-BSA cross-linking agent is immobilized on the nitrocellulose membrane as the T line, while mouse anti-AFB1 antibody is combined with UCNPs covalently. For positive detection, UCNPs-anti-AFB1 antibody is combined with AFB1 in samples and cannot be further captured by AFB1-BSA crosslinking agent on the nitrocellulose membrane, leading to the

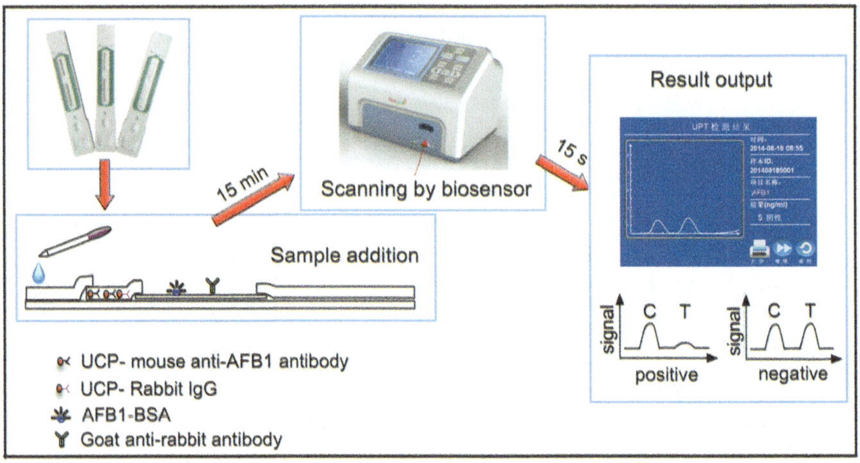

Fig. 16.2 Schematic for detection of AFB1 by UPT-POCT based on competitive mode

decrease of signals. Goat anti-rabbit antibody at the C line immobilized on the nitrocellulose membrane can capture UCNPs-rabbit IgG forming a stable control signal. T/C values are negatively proportional to the concentrations of AFB1 in samples.

The UPT-POCT assay for multiplex detection is based on a 10-channel UPT-LF disc or multiple T lines for one strip. The UPT-LF disc is composed of ten one-target strips, therefore ten targets can be detected from the loading of one sample. For instance, a UPT-LF disc prepared with ten proteins of *Y. pestis* based on the double-antigen sandwich mode has been realized for the detection of antibodies against *Y. pestis*, providing a clue for seeking diagnosis biomarker of *Y. pestis* (Hong et al. 2010). A strip with multiple T lines can detect multiple targets, such as the UPT-LF assay developed for simultaneous detection of *V. cholerae* serogroup O1 and O139 (Hao et al. 2017).

16.5.2 *Performance Evaluation*

Sensitivity and specificity are crucial to the performance evaluation for a detection method, and the evaluations of UPT-POCT for detection of different bioterrorism agents are shown in Table 16.1. The sensitivity of UPT-POCT for bacterial bioterrorism agents can reach 10^3 CFU/ml (namely 10 CFU for each test based on ten-fold sample dilution and 100 μL of sample loading volume), and that for toxin can reach 0.03 ng/ml. The quantitative range covers four to five orders of magnitude, even six for detection of *B. anthracis* spore (Zhang et al. 2014). The coefficients of variation for detection are all below 15%. UPT-POCT shows excellent specificity to the bioterrorism agents that share genetic relatedness, similar transmission routes, or similar structure with the targets.

Table 16.1 Sensitivity and specificity of UPT-POCT for bioterrorism agents

Classification	Detection target	Sensitivity	Quantitative range	Specificity
Bacteria	*Y. pestis* (Yan et al. 2006; Zhang et al. 2014)	10^4 cfu/ml	10^4–10^8 cfu/ml	Specific to *Y. aldovae*, *Y. enterocolitica*, *Y. intermedia*, *Y. kristensenii*, *Y. pseudotuberculosis*, *Y. rohdei*, *Y. ruckeri*, *B. anthracis* spore, *B. melitensis* M55009, *E. coli*, and *S. choleraesuis*
	B. anthracis spore (Li et al. 2006; Zhang et al. 2014)	10^5 cfu/ml	10^5–10^{10} cfu/ml	Specific to *B. antrophaeus* spore, *B. thuringiensis* spore, *B. megaterium*, *B. mycoides*, *B. melitensis* M55009, and *Y. pestis*; Serious cross reaction with some isolates of *B. cereus* spore and *B. subtilis* spore because of the high similarity between spore structures
	Brucella spp. (Qu et al. 2009; Zhang et al. 2014)	10^6 cfu/ml	10^6–10^9 cfu/ml	Specific to *E. coli* O157:H7, *Salmonella* spp. (including *S. choleraesuis*, *S. enteritidis*, *S. paratyphi* A, *S. paratyphi* B, *S. paratyphi* C, *S. typhi*, and *S. typhimurium*), *V. cholerae* O1, *V. cholerae* O139, *Y. pseudotuberculosis*, *Y. enterocolitica*, *B. anthracis* spore, and *Y. pesits*

(continued)

Table 16.1 (continued)

Classification	Detection target	Sensitivity	Quantitative range	Specificity
	B. pseudomallei (Hua et al. 2015a)	10^4 cfu/ml	10^4–10^7 cfu/ml	Specific to the stains that shared close genetic relatedness: *P. aeruginosa* and *Burkholderi* spp. (including *B. mallei*, *B. cocovenenans*, *B. thailandensis*, *B. glathei*、*B. gladiol*, *B. vietnamiensis*, *B. cepacia*, *B. phenazinium*); Specific to *B. anthracis* spore, *B. melitensis* M55009, *F. tularensis*, *Y. pestis*, *E. coli* O157:H7, *L. innocua*, *L. monocytogenes*, *Salmonella* spp. (including *S. choleraesuis*, *S. enteritidis*, *S. paratyphi* A, *S. paratyphi* B, *S. paratyphi* C, *S. typhi*, and *S. typhimurium*), *S. dysenteriae*, *V. cholerae* O1 and O139, and *V. parahaemolyticus*
	F. tularensis (Hua et al. 2015b)	10^4 cfu/ml	10^4–10^8 cfu/ml	Specific to *B. anthracis* spore, *B. melitensis* M55009, *B. pseudomallei*, *Y. pestis*, *E.coli* O157:H7, *L. innocua*, *L. monocytogenes*, *Salmonella* spp. (including *S. choleraesuis*, *S. enteritidis*, *S. paratyphi* A, *S. paratyphi* B, *S. paratyphi* C, *S.typhi*, and *S. typhimurium*), *V. cholerae* O1, *V. cholerae* O139, and *V. parahaemolyticus*; Slight cross reaction with *S. dysenteriae*

(continued)

Table 16.1 (continued)

Classification	Detection target	Sensitivity	Quantitative range	Specificity
	V. cholerae (Hao et al. 2017)	10^4 cfu/ml	10^4–10^8 cfu/ml	Specific to *V. fluvialis*, *V. metschnikovii*, *V. minicus*, *V. Vulnificus*, *V. parahaemolyticus*, *A. hydrophila*, *E. coli*, *salmonella* spp., and *S. flexneri*
	E. coli O157 (Wang et al. 2007)	10^3 cfu/ml	–	Specific to *E. coli* K0111:K74, *Citrobacter freundii*, *S. paratyphi* A, *S. paratyphi* B, *S. enteritidis*, *S. boydii*, *S. glexneri*, *Proteus*, *Enteroaerogen*, *Citrobacter*, *Serratia*, *Staphylococcus*, *V. parahemolyticus*, and *Listeria monocytogenes*
Toxin	Ricin (Wang et al. 2016)	0.5 ng/ml	0.5–1000 ng/ml	Specific to Shiga toxin I, Shiga toxin II, aflatoxin M1, aflatoxin B1, ochratoxin, and abrin
	Abrin (Liu et al. 2016)	0.1 ng/ml	0.1–1000 ng/ml	Specific to shiga toxin I, Shiga toxin II, aflatoxin M1, aflatoxin B1, ochratoxin, ricin, *S. aureus* enterotoxin B, and botulinum toxin
	Aflatoxin B1 (Zhao et al. 2016)	0.03 ng/ml	0.03–1000 ng/ml	Specific to aflatoxin M1, ochratoxin, abrin, ricin, shiga toxin I, shiga toxin II, *S. aureus* enterotoxin B, and botulinum toxin
	Aflatoxin M1 (Liu et al. 2014)	0.1 μg/kg milk powder; 0.3 μg/L milk	0.1–0.7 μg/kg milk powder; 0.3–0.7 μg/L milk	–
Virus	Antibody against hepatitis virus	10 mIU/mL	20–900 mIU/mL	–

The tolerance to biochemical agents (Table 16.2) and operation error of UPT-POCT for the detection of five bacterial bioterrorism agents has been evaluated (Zhang et al. 2014; Hua et al. 2015a, b). The high concentration of agent that the detection method can tolerate is defined as the tolerance limit. The tolerance limits of UPT-POCT for pH, ionic strength, viscosity, and bio-macromolecule concentration can reach pH 1 –13, ≥ 4 mol/L of NaCl and KCl solution, ≤ 100 mg/ml PEG 2000, ≥ 20% glycerol, ≥ 400 mg/ml of BSA and ≥ 80 mg/ml of casein, respectively. At some tolerance limits, the sensitivity could be improved by one order of magnitude. The operation error, including the volume variation of the sample (from −50 to 200%), sample treating buffer (from −22 to 44%), and loading mixture (from −30 to 30%), has little influence on the sensitivity and specificity of UPT-POCT.

16.5.3 Field Evaluation

UPT-POCT shows excellent performance for detection of bacterial bioterrorism agents in simulated samples, such as food, powder, and viscera (Zhang et al. 2014; Hua et al. 2015a, b) (Table 16.3), and it can also effectively detect abrin and afaltoxin B1 in food and grains (Liu et al. 2016; Zhao et al. 2016) (Table 16.4). The highest tolerance limit of UPT-POCT to simulated samples could reach 400 mg/ml.

For detection of *V. cholerae* in 102 field water samples obtained from sample collection sites in Guangzhou city (China), UPT-POCT is more sensitive than the isolation-culture method and colloidal gold immnochromatography assay, and its sensitivity could match that of real-time fluorescent PCR with fewer false positive results (Hao et al. 2017).

The sample pre-treatment for detection by UPT-POCT is merely grinding, or supernatant extraction by centrifugation after grinding and shaking on a vortex shaker for 15 min (or ultrasonicating for 10 min), and then the sample can be directly mixed with sample-treating buffer for detection. The tolerance of UPT-POCT to sample is based on the physical and luminescence stability of UCNPs, the stable covalent interaction between UCNPs and antibody, as well as the strong buffering capacity of the detection system.

16.6 Application

16.6.1 Current Detection Methods

Detection methods for bioterrorism agents include the isolation-culture or animal inoculation method, biochemical method, nucleic acid method, and immunization method.

Table 16.2 Tolerance limits of UPT-POCT for biochemical reagents in the detection of five bacterial bioterrorism agents

Interference Factor		Unit	Detection target				
			B. anthracis Spore (Zhang et al. 2014)	*Brucella* spp. (Zhang et al. 2014)	*Y. pestis* (Zhang et al. 2014)	*B. pseudomallei* (Hua et al. 2015a)	*F. tularensis* (Hua et al. 2015b)
pH value	HCl	mol/L	≤ 0.001 (pH 3)	≤ 0.01 (pH 2)**	≤ 0.01 (pH 2)	≤ 0.1 (pH 1)	≤ 0.01 (pH 2)
	NaOH	mol/L	≤ 0.01 (pH 12)	≤ 0.01 (pH 12)**	≤ 0.001 (pH 11)	≤ 0.01 (pH 2)	≥ 0.1 (pH 13)
Ion strength	NaCl +KCl	mol/L	≤ 0.25	≥ 4**	≤ 2	≤ 2	≥ 2
Viscosity	PEG20000	mg/ml	≤ 12.5	≤ 25**	≤ 12.5	≤ 100	≤ 50
	Glycerol	% (V/V)	≥ 20%	< 5%	≤ 5%	≤ 20%	≥ 20%
Bio-macromolecule	BSA	mg/ml	≤ 100	≤ 200**	≤ 100	≥ 400	≥ 400
	Casein	mg/ml	≤ 5	≥ 80**	≤ 40	≥ 80	≥ 80

**The sensitivity of UPT-LF strip improved by one order of magnitude at that tolerance limit

Table 16.3 Tolerance limits of UPT-POCT for simulated samples in the detection of five bacterial bioterrorism agents

Simulated sample (mg/ml)		Detection target				
		B. anthracis spore (Zhang et al. 2014)	*Brucella* spp. (Zhang et al. 2014)	*Y. pestis* (Zhang et al. 2014)	*B. pseudomallei* (Hua et al. 2015a)	*F. tularensis* (Hua et al. 2015b)
Power	Flour	≤ 100	≥ 200	≤ 50	≥ 400	≥ 200
	Fruit juice	≤ 100	≤ 50**	≤ 50	≥ 400	≥ 200
	Gourmet powder	≥ 400	≥ 400**	≤ 50	≤ 200	≥ 200
	Milk powder	≤ 25	≤ 200**	≥ 400	≥ 400	≤ 50
	Putty powder	≥ 200	≥ 200**	≤ 50	≤ 200	≥ 200
	Soil	≥ 400	≥ 400**	≥ 400	≤ 200**	≤ 100
	Sucrose	≤ 100	≥ 400**	≥ 400	≤ 200	≥ 200
Viscera (mouse)	Fresh heart	≥ 800	≥ 800	≥ 800	≥ 400	≥ 400
	Fresh liver	≤ 50	≤ 200**	≤ 50	≥ 400**	≥ 400
	Fresh lung	≤ 400	≥ 800	≤ 100	≥ 400	≥ 400
	Fresh spleen	≤ 200	≥ 400	≤ 100	≥ 400**	≥ 400
	Decomposed heart	≤ 100	≥ 400**	≤ 100	≥ 400	≥ 400
	Decomposed liver	≤ 50	≤ 100**	≤ 200	≥ 400	≥ 400
	Decomposed lung	≤ 100	≤ 200**	≤ 200	≥ 400	≥ 400
	Decomposed spleen	≤ 100	≤ 100**	≤ 100	≥ 400**	≥ 400

** The sensitivity of UPT-LF strip improved by one order of magnitude at that tolerance limit

Table 16.4 Tolerance limits of UPT-LF assay for simulated samples for two toxin bioterrorism agents

Abrin (Liu et al. 2016)		Aflatoxin B1 (Zhao et al. 2016)	
Simulated sample	Tolerance limit (detection limit)	Simulated sample	Tolerance limit (detection limit)
Cookie	30 mg/ml (3.33 ng/g)	Peanut	300 mg/ml (0.1 ng/g)
Soybean	50 mg/ml (2 ng/g)	Runner bean	200 mg/ml (0.15 ng/g)
Sausage	200 mg/ml (0.5 ng/g)	Common bean	200 mg/ml (0.15 ng/g)
Cashew	100 mg/ml (1 ng/g)	Semen phaseoli	200 mg/ml (0.15 ng/g)
Milk powder	80 mg/ml (1.25 ng/g)	Rice	200 mg/ml (0.15 ng/g)
Flour	40 mg/ml (2.5 ng/g)	Barley	200 mg/ml (0.15 ng/g)
Sugar	10 mg/ml (10 ng/g)	Mung bean	200 mg/ml (0.15 ng/g)
Monosodium Glutamate	12.5 mg/ml (8 ng/g)	Corn	100 mg/ml (0.30 ng/g)
Water	2.5:5 (0.3 ng/ml)	Adzuki bean	100 mg/ml (0.30 ng/g)
Soft drink	2:5 (0.35 ng/ml)	Soybean	100 mg/ml (0.30 ng/g)
Juice	2:5 (0.35 ng/ml)	Black rice	50 mg/ml (0.60 ng/g)
Beer	1.5:5 (0.43 ng/ml)	Broomcorn	50 mg/ml (0.60 ng/g)
–	–	Oats	50 mg/ml (0.60 ng/g)
–	–	Brown rice	100 mg/ml (5 ng/g)
–	–	Coix seed	200 mg/ml (2.5 ng/g)

16.6.1.1 Isolation-Culture or Animal Inoculation Method

Microorganism bioterrorism agents can be identified by the isolation-culture method or inoculation of animals, and toxin can also be injected into animals for species identification. Selective culture medium, as well as inoculation and dissection of susceptible animals, are both common experiments. The culture methods for the various bioterrorism agents are different. (1) Bacteria can be identified by selection in selective medium, for example, alkaline peptone water is the medium for selective culture of *V. cholerea*, while *Y. pestis* can be identified by culturing with bacteriophage lysis. (2) Isolation of viruses by cell culture is the main method for virus detection because viruses can only survive in live cells. After virus infection the mutated cell can be directly detected by microscopy, or observed through the change in pH of the medium, hemadsorption or hemagglutination. Culture in chick embryo is also common for viruses such as influenza virus. Animal incubations are better than cell incubation for some viruses, for example, inoculation of mice is the best culture method for Rabies virus. (3) Rickettsia bioterrorism-associated agents are cultured and isolated by guinea pig and chick embryo. (4) The types of toxin can be identified by lethality or the animal response after inoculation of susceptible animals or cells, such as vomiting and diarrhea caused by *S. aureus* enterotoxin B. The susceptible animals for toxins are different, for instance, mouse and cat are sensitive for botulinum toxin and *S. aureus* enterotoxin, respectively.

The isolation-culture method is the basic detection method—even gold standard detection method— for bacterial bioterrorism agents. However, it must be conducted by professionals in particular institutes equipped with biosafety facilities, and bioterrorism agents can be easily transmitted owing to improper operation or defense. Because of its low sensitivity, the isolation-culture method is often combined with other methods to improve the accuracy of detection, such as the bacteriophage lysis method (Zhao et al. 2013). For example, the bacteriophage lysis method for identification of *Y. pestis* is realized through adding bacteriophage into cultured bacteria at 18–20 °C. The low sensitivity and specificity of the animal injection method for detection of toxins are caused by the diversity of toxins and individual differences between animals.

16.6.1.2 Biochemical Method

The biochemical method is based on the properties or metabolism characteristics of the microorganism or toxin, and is in fact a detection system because the species cannot be determined by one property, such as the systemic biochemical detection according to diagnostic criteria for cholera (WS 289-2008, healthy industry standards of the People's Republic of China). The poor sensitivity and specificity of the biochemical method, as well as the complicated operation, are a result of the property similarities between microorganisms or biological substances.

16.6.1.3 Nucleic Acid Method

The nucleic acid method is based on the principle of DNA replication in vitro, such as the polymerase chain reaction (PCR) and Loop-mediated isothermal amplification (LAMP).

PCR is a laboratory detection method that parallels the DNA replication process in vivo. The single strand DNA template is formed at 95 °C, and then it can match with a primer at annealing temperature (about 55 °C) based on their complementary sequences, subsequently a new complementary DNA strand can be obtained using dNTP as a reactive material catalyzed by Taq DNA polymerase. These strands can be further used as a template for the next cycle, therefore the target DNA can be multiplied millions of times by dozens of cycles. The products can be determined by DNA electrophoresis for common PCR, while for real-time quantitative PCR the amplification process can be monitored by the change of fluorescence signals. In addition to DNA as a template, RNA could also be used for amplification, and this is realized using the reverse transcriptional PCR (RT-PCR) method.

The target genes of PCR for *Y. pestis* include the *caf*1 gene that encodes F1 antigen, *ymt* gene that encodes murine toxin, *pla* gene that encodes plasminogen activator, *hms* gene that is related to pigmentation, and specific segment 3a in chromosome (Qu et al. 2010). Segment 3a is the main target gene because it is not as easily lost as the genes of plasmid such as *pla* and *caf*1. The target genes of PCR for *F. tularensis* include *fop*A gene (AY579741) that encodes outer membrane protein, and the *akr* gene (AM286280, 959924-960988) in chromosomes that encodes Aldo/ketoreductase. The genes at two specific toxin plasmid for *B. anthracis*, including *cya*, *lef*, *pagA* at plasmid pXO1 and *capA*, *capB*, *capC* at plasmid PXO2, are often used for species identification (Koehler 2009), while further identification by the detection of the genes in chromosomes, such as GS sequences, is necessary because of the inaccuracy of detection results caused by the deletion or change of plasmids.

The PCR method has higher sensitivity than other current methods, but its application on site as a POCT method is limited by its dependency on expensive equipment, sample pre-treatment (particularly difficult for DNA extraction from complicated samples) and professional operation, as well as its high false positive results. For detection of bioterrorism agents, the special biosecurity facilities for DNA extraction (even in an equipped laboratory the infection and contamination in the DNA extraction procedure must be paid particular attention) were the major obstacle for utilization of PCR on site. Currently, an instrument based on PCR, called a Film-array (Seiner et al. 2013), demonstrates a promising prospect for application of PCT in the field by integration of the sample treatment, amplification, and result analysis in airtight system. However, the complexity of pre-treatment of complicated samples in routine surveillance and public health emergencies is still a significant limitation of application of PCR on site.

Loop-mediated isothermal amplification (LAMP) is a new method for gene diagnosis invented by a Japanese professor, in which a DNA strand that is complementary to template DNA can be synthesized at a determined temperature through strand displacement reaction. Using four primers designed according the six segments of the target DNA, LAMP detection can be realized through one procedure following mixing of the template, primer, strand displacement type DNA polymerase and other substrates. The pyrophosphate isolated from dNTPs in the DNA synthesis process can react with Mg^{2+} ions, resulting in the formation of a white precipitate. LAMP has been developed for detection of some bioterrorism agents, such as *Y. pestis* (Feng et al. 2017) and *B. anthracis* (Qiao et al. 2007), and it is very suitable for rapid detection on site because only a thermostat is needed and the results can be observed by the naked eye. However, it is not suitable for long strand target DNA, because sequences longer than 500 bp are difficult to amplify based on the strand replacement reaction. Because LAMP is an amplification reaction similar to PCR, a high frequency of false positive results are often generated by LAMP because of contamination.

16.6.1.4 Immunization Method

The immunization method is based on the reaction between antigens and antibodies, such as enzyme linked immunosorbent assay (ELISA), immunochromatography, immunodiffusion, and immunoprecipitation. Immunodiffusion and immunoprecipitation were developed in the 1950s, and are not widely used at present owing to poor sensitivity. ELISA is a routine laboratory method with excellent sensitivity and specificity owing to signal amplification by enzymes and several rinse procedures. However, a high number of rinse procedures increases the complexity of the operation and the operation error, as well as the potential for spread of bioterrorism-associated agents. Immunochromatography was described in Sect. 16.4.2.

16.6.2 The Merit of UPT-POCT and Its Application

Many UPT-LF assays have been developed for detection of several bioterrorism-associated agents, including detection of bacteria (*Y. pestis*, *Brucella*, *B. anthracis* spore, *F. tularensis*, *B. pseudomallei*, *V. cholorae*, *E. coli* O157:H7, antibody against *Y. pestis*), viruses (hepatitis B surface antibody) and toxins (abrin, ricin, aflatoxin B1, aflatoxin M1 and T-2 toxin). UPT-POCT has been provided at many centers for disease control and prevention, as well as entry-exit inspection and quarantine bureaus, and it also provides technology support for biosecurity at large events, such as the Games of the 2008 Olympiad in Beijing, Shanghai world Exposition, and Asia Games in Guangzhou. In 2011, UPT-POCT was integrated as a mobile biological rapid detection instrument in the Stand for Construction of City fire Station (152-2011, issued by the Ministry of Housing and Urban-Rural Development of China and the National Development and Reform Commission of China).

UPT-POCT is suitable for detection of bioterrorism agents on site as a POCT method because of the integration of the benefits of immunochromatography, up-converting phosphor particles, and portable biosensing.

First, UPT-POCT can give reliable detection results. Given the extremely low pathogenic dose and high social perniciousness of bioterrorism agents, particularly the high transmission capacity of microorganism agents, detection sensitivity and reliability are important for detection methods. Compared with the isolation-culture method, biochemical method, and colloidal gold method, UPT-POCT can realize sensitive and quantitative detection based on immunological interactions, and shows excellent performance that is comparable with that of real-time quantitative PCR in some applications (Hao et al. 2017), resulting from the merits of the immunological recognition mode, up-converting phosphor, and biosensor. First, immunological recognition between antigens and antibodies is highly sensitive and specific. In addition, the infrared excitation light for up-converting phosphor particles avoids the interference from other biomaterial in the samples, resulting in a more efficient signal isolation rate than other luminous detection methods. The efficient signal extraction and quantitation calculation by the biosensors also facilitate the recognition of the weak positive signal, which is superior to observation with the naked eye for colloidal-gold immunochromatography assays.

UPT-POCT can be tolerant to a great diversity of complex samples in the bioterrorism and biosecurity fields (such as meat, decomposed viscera, and flour). Many detection methods are limited by the complicated pre-treatment of samples. For example, repeated selection and identification is required for the isolation-culture method, and the complex composition of samples can easily influence the result of biochemical detection. The pretreatment of complex samples is also a significant challenge for laboratory methods, such as the extraction of DNA in complicated samples in PCR that could seriously influence the detection rate. In detection of many bioterrorism agents, UPT-POCT shows robust performance for various samples with simple pretreatments, such as grinding, or supernatant extraction by centrifugation after grinding and shaking (ultrasonic). This robust performance derives from the physical stability, luminous stability, and up-converting capacity of UCNPs, the solid covalent combination between the UCNPs and antibody, and the excellent buffering capacity of the detection system.

UPT-POCT is also safer than other detection methods. There are some operations that are unfavorable for the control of bioterrorism agents during detection using other methods, such as repeated proliferation for the isolation-cultured method, various rinses for ELISA, complex pre-treatment of sample for the PCR method, and multiple tests for bio-chemical detection. Compared with these methods, the simple sample-treatment process based on its high tolerance, and the simple sample-loading manner of UPT-POCT, reduces the potential for the spread of bioterrorism agents in the detection process.

The short acquisition time for UPT-POCT facilitates rapid response to disease outbreaks in surveillance and public health emergencies, and is derived from the 15 min reaction process of the immunochromatography detection mode and the simple sample pretreatment. As a quantitative detection method, the detection time needed for PCR is more than that for UPT-POCT owing to the time required for DNA extraction.

Portability is the main obstacle to many laboratory detection methods for application on site, for example the expensive and cumbersome equipment for the PCR method. The portability of UPT-POCT derives from the small size of the strips and biosensors. In addition, the biosensor could work with a standard mains electricity source or battery.

For UPT-POCT, reliable results can be obtained by nonprofessionals owing to the simple nature of UPT-POCT operation and sample pre-treatment, making it possible to treat incidents rapidly for surveillance and in public health emergencies.

Acknowledgements We are grateful to Jin Wang, Xiangxiu Qin, Lili Zhang, Jianhua Li, and Yan Wang from the Institute of Microbiology and Epidemiology for their devotion to antibody preparation. The creative research of UCNP modification, UPT biosensors, and UPT-LF strips was accomplished by Yan Zheng from Shanghai Kerune Phosphor Technology Co. Ltd. (Shanghai, China), Huijie Huang and Lihua Huang from Shanghai Institute of Optics and Fine Mechanics, Chinese academy of Sciences, as well as Ruifu Yang and Lei Zhou from the Institute of Microbiology and Epidemiology. In addition, Changqing Lin from Beijing Hotgen Biological Technology Co. Ltd. (Beijing, China) made a significant contribution to the industrialization of UPT program.

This study was supported by the National High Technology Research and Development Program of China (No. 2013AA032205), Major National Science and Technology Programs of China (No. 2011ZX10004 and 2012ZX1004801), Beijing Nova Programme (No. Z151100000315086), and grant from the Key Laboratory of Food Safety Risk Assessment of Ministry of Health, China National Center for Food Safety Risk Assessment (No. 2015K03).

References

Agar SL, Sha J, Foltz SM, et al. Characterization of the rat pneumonic plague model: infection kinetics following aerosolization of *Yersinia pestis* CO92. Microbes Infect. 2009;11(2):205–14.

Begier EM, Asiki G, Anywaine Z, et al. Pneumonic plague cluster, Uganda, 2004. Emerg Infect Dis. 2006;12(3):460–7.

Chitadze N, Kuchuloria T, Clark DV, et al. Water-borne outbreak of oropharyngeal and glandular tularemia in Georgia: investigation and follow-up. Infection. 2009;37(6):514–21.

Draper AD, Mayo M, Harrington G, et al. Association of the melioidosis agent *Burkholderia pseudomallei* with water parameters in rural water supplies in Northern Australia. Appl Environ Microbiol. 2010;76(15):5305–7.

Estill CF. recovery efficiency and limit of detection of aerosolized Bacillus anthracis Sterne from environmental surface samples. Appl Environ Microbiol. 2009;75:4297–306.

Feng N, Zhou Y, Fan Y, et al. *Yersinia pestis* detection by loop-mediated isothermal amplification combined with magnetic bead capture of DNA. Braz J Microbiol. 2017.

Finelli L, Swerdlow D, Mertz K, et al. Outbreak of cholera associated with crab brought from an area with epidemic disease. J Infect Dis. 1992;166(6):1433–5.

Gouvras G. The far-reaching impact of bioterrorism: what the European Union is doing regarding deliberate releases of biological/chemical agents based on the events in the United States. IEEE Eng Med Biol Mag. 2002;21(5):112–5.

Hao M, Zhang P, Li B, et al. Development and evaluation of an up-converting phosphor technology-based lateral flow assay for the rapid, simultaneous detection of *Vibrio cholerae* serogroups O1 and O139. PLoS ONE. 2017;12(6):e0179937.

Hill VR, Cohen N, Kahler AM, et al. Toxigenic *Vibrio cholerae* O1 in water and seafood, Haiti. Emerg Infect Dis. 2011;17(11):2147–50.

Hong W, Huang L, Wang H, et al. Development of an up-converting phosphor technology-based 10-channel lateral flow assay for profiling antibodies against *Yersinia pestis*. J Microbiol Methods. 2010;83(2):133–40.

Hua F, Zhang P, Wang X, et al. Development and systematical evaluation of an up-converting phosphor technology based lateral flow assay for quantitative detection of *Burkholderia pseudomallei*. Clin J Prev Med. 2015a;49(2):166–71.

Hua F, Zhang P, Zhang F, et al. Development and evaluation of an up-converting phosphor technology-based lateral flow assay for rapid detection of *Francisella tularensis*. Sci Rep. 2015b;5:17178.

Huq A, Colwell RR, Chowdhury MA, et al. Coexistence of *Vibrio cholerae* O1 and O139 Bengal in plankton in Bangladesh. Lancet. 1995;345(8959):1249.

Koehler TM. *Bacillus anthracis* physiology and genetics. Mol Aspects Med. 2009;30(6):386–96.

Li L, Zhou L, Yu Y, et al. Development of up-converting phosphor technology-based lateral-flow assay for rapidly quantitative detection of hepatitis B surface antibody. Diagn Microbiol Infect Dis. 2009;63(2):65–172.

Li W, Zhou L, Wang J, et al. Development of up–converting phosphor immunochromatography for fast and quantitive detection of *Bacillus anthracis* spore. Clin J Microbiol Immunol. 2006;26(8):761–4.

Liu X, Wang LP, Zhou L, et al. Rapid detection of aflatoxin M1 in milk powder and milk based on up-converting phosphor technology. Mil Med Sci. 2014;38(11):850–4.

Liu X, Zhao Y, Sun C, et al. Rapid detection of abrin in foods with an up-converting phosphor technology-based lateral flow assay. Sci Rep. 2016;6:34926.

Lukaszewski RA, Kenny DJ, Taylor R, et al. Pathogenesis of *Yersinia pestis* infection in BALB/c mice: effects on host macrophages and neutrophils. Infect Immun. 2005;73(11):7142–50.

Porche DJ. Biological and chemical bioterrorism agents. J Assoc Nurses AIDS Care. 2002;13(5):57–64.

Prentice MB, Rahalison L. Plague. Lancet. 2007;369(9568):1196–207.

Qiao YM, Guo YC, Zhang XE, et al. Loop-mediated isothermal amplification for rapid detection of *Bacillus anthracis* spores. Biotechnol Lett. 2007;29(12):1939.

Qu Q, Zhu Z, Wang Y, et al. Rapid and quantitative detection of *Brucella* by up-converting phosphor technology-based lateral-flow assay. J Microbiol Methods. 2009;79(1):121–3.

Qu S, Shi Q, Zhou L, et al. Ambient stable quantitative PCR reagents for the detection of *Yersinia pestis*. PLoS Negl Trop Dis. 2010;4(3):e629.

Seiner DR, Colburn HA, Baird C, et al. Evaluation of the FilmArray(R) system for detection of *Bacillus anthracis*, *Francisella tularensis* and *Yersinia pestis*. J Appl Microbiol. 2013;114(4):992–1000.

Simsek H, Taner M, Karadenizli A, et al. Identification of *Francisella tularensis* by both culture and real-time TaqMan PCR methods from environmental water specimens in outbreak areas where tularemia cases were not previously reported. Eur J Clin Microbiol Infect Dis. 2012;31(9):2353–7.

Stralin K, Eliasson H, Back E. An outbreak of primary pneumonic tularemia. N Engl J Med. 2002;346(13):1027–9.

Wang J, Zhou L, Wei LI, et al. Development of UCP-immunochromatography Test for Rapid Detection of *E. coli* O157. J Hyg Res. 2007;35(4): 439.

Wang JL. Toxin warfare agents: recognition molecules and drugs for control. Mil Med Sci. 2011;35(8):561–5.

Wang X, Lei Z, chongyun S, et al. Preparation of monoclonal antibodies against ricin toxin and development of up-converting phosphor technology-based lateral flow assay for its quantitative detection. Mil Med Sci. 2016;40(8): 676–9.

Yan Z, Zhou L, Zhao Y, et al. Rapid quantitative detection of *Yersinia pestis* by lateral-flow immunoassay and up-converting phosphor technology-based biosensor. Sens Actuators B Chem. 2006;119(2):656–63.

Yang R. Fang sheng wu wei hai yi xue. Military Medical Science Press;2008. p. 9–13.

Zhang P, Liu X, Wang C, et al. Evaluation of up-converting phosphor technology-based lateral flow strips for rapid detection of *Bacillus anthracis* Spore, *Brucella* spp., and *Yersinia pestis*. PLoS One. 2014;9(8):e105305.

Zhao X, Cui Y, Yan Y, et al. Outer membrane proteins ail and OmpF of *Yersinia pestis* are involved in the adsorption of T7-related bacteriophage Yep-phi. J Virol. 2013;87(22):12260–9.

Zhao Y, Liu X, Wang X, et al. Development and evaluation of an up-converting phosphor technology-based lateral flow assay for rapid and quantitative detection of aflatoxin B1 in crops. Talanta. 2016;161:297–303.

Chapter 17
Application of UCNPs in Bio-imaging and Treatment

Yong Zhao

Abstract Upconversion nanoparticles (UCNPs) are ideal fluorescent probes for biomedical applications owing to their good characteristics, including superior photostability, deep light penetration, low background auto-fluorescence and good biocompatibility. This chapter focuses on the recent developments of UCNPs in bioimaging and tumor therapy applications, as well as gives an analysis of the advantages of UCNPs over the conventional fluorescent materials and the biocompatibility and toxicity of UCNPs. Finally, the chapter discuss the challenges in the development of UCNPs in biomedical application.

Keywords UCNPs · Bio-imaging · Tumor therapy

In biomedical research, fluorescent probes play an important role in the exploration of life processes. Currently, there are three main types of fluorescent probe: organic fluorescent dyes, semiconductor quantum dots, and upconversion nanoparticles (UCNPs). Among them, organic fluorescent dyes are widely used, however their photostability is poor, and the emitted light significantly attenuates under extended light exposure, making them unsuitable for continuous monitoring in vivo. Although semiconductor quantum dots have good photostability, a narrow emission spectrum, and high quantum yield, the potential toxicity of heavy metal elements limits their biological application. In addition, both organic fluorescent dyes and quantum dots are down-converting luminescent materials, and biological tissues can absorb the excitation light and interfere with the probe signal. In comparison, UCNPs luminescence uses near-infrared lasers as the excitation light source, which offer many unique advantages, such as excellent photostability, deep light penetration, minimal photo damage to living organisms, and low background auto-fluorescence. UCNPs are therefore considered ideal materials for biomedical applications.

Y. Zhao (✉)
Beijing Key Laboratory of POCT for Bioemergency and Clinic (No. BZ0329), Beijing 100071, China
e-mail: zhaoyong179@139.com

Beijing Institute of Microbiology and Epidemiology, No. 20 Dongdajie, Beijing 100071, Fengtai, People's Republic of China

R. Yang (ed.), *Principles and Applications of Up-converting Phosphor Technology*,
https://doi.org/10.1007/978-981-32-9279-6_17

In recent years, there has been a continuous push to develop UCNPs-based technology for bio-imaging applications. For example, UCNPs-based multimode imaging technology, which combines UCNPs luminescence imaging with other imaging techniques such as magnetic resonance imaging (MRI) and computer tomography (CT), has gradually gained research interest. In addition, with superior photostability, long lifetimes, narrow emission spectra, and good biocompatibility, UCNPs materials are also potential drug carriers, and could be coupled with photosensitizer for application in tumor therapy. This chapter focuses on the recent developments of UCNPs in bio-imaging applications, the exploration of UCNPs as multifunctional nanoscale carriers for tumor treatment, and the challenges and opportunities for UCNPs in biomedical application.

17.1 Application of UCNPs in Bio-imaging

17.1.1 Deep Tissue Imaging

UCNPs are ideal materials for deep tissue imaging in vivo because of their strong penetration ability, minimal photo damage to biological tissues, and low background of auto-fluorescence.

In 2006, Lim reported the first application of UCNPs (Y_2O_3:Yb/Er, particle size 50–150 nm) to in vivo imaging of nematodes (Lim et al. 2006). The distribution of UCNPs in nematodes could be clearly observed under the excitation of a 980 nm laser. It was also reported that polyacrylic acid-modified UCNPs (PAA-$NaLuF_4$:Yb, Tm) could be used for high quality fluorescence imaging of normal black rats and rabbits (Yang et al. 2012). With the rapid development of nanotechnology, the application of UCNPs for in vivo imaging has been further developed in recent years. Abdul Jalil and Zhang (2008) synthesized silicon coated UCNPs and injected them into Balb/C mice via the tail vein. Subsequently, strong up-converted luminescence signals were found in the ear vessels of the mice. Nyk et al. (2008) applied UCNPs ($NaYF_4$:Yb, Tm) with particle size of 20–30 nm to mice via tail vein. Under the excitation of near-infrared light (980 nm), it was found UCNPs were mainly concentrated in the liver. The imaging had a high signal-to-noise ratio even at a penetration depth of 20 mm, highlighting the advantages of UCNPs in bioimaging.

17.1.2 Targeted Imaging of Tumor Cells

Targeted imaging of tumor cells plays an important role in tumor diagnosis and prognosis. Owing to their unique optical properties, researchers have carried out a great deal of work on the application of UCNPs to the targeted imaging of tumor cells. In 2009, Li and colleagues (Xiong et al. 2009a; Yu et al. 2009) reported the first

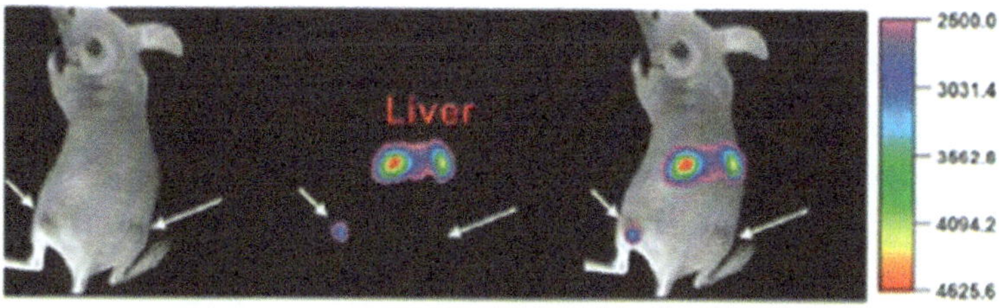

Fig. 17.1 Upconversion luminescent imaging of U87MG tumor cells in vivo (Xiong et al. 2009b)

application of UCNPs to tumor imaging. In this work, UCNPs ($NaYF_4$:Yb, Er) with aminated surfaces were first conjugated to folic acid molecules (FA), then the UCNPs-FA complexes were cultured with FA receptor positive and FA receptor negative HeLa cells. Under 980 nm light irradiation, significant upconversion luminescence signals were observed for the FA receptor positive HeLa cell sample, whereas the signal was not observed for the FA receptor negative HeLa cells. Furthermore, UCNPs-FA complexes were injected into Balb/C mice (carrying HeLa cells) through the tail vein. After 24 h, upconversion luminescence signals were observed in the tumor region of the mice, which demonstrated the potential for UCNPs to be applied in targeted tumor cell imaging.

In other work (Xiong et al. 2009b), Li and coworkers applied UCNPs for targeted imaging of U87MG tumor cells in a mouse model through modification the surface of the UCNPs with RGD peptides (Fig. 17.1), which have a high affinity for $\alpha v\beta 3$ integrin receptor on the cell membrane of U87MG tumor cells. Results of the tissue slice imaging data and ROI (region of interest) analysis showed that UCNPs have a high signal-to-noise ratio and imaging depth. Moreover, considering the high affinity of neurotoxins for tumor cells, some researchers have linked neurotoxins to the surface of UCNPs to achieve targeted imaging of tumor cells (Yu et al. 2010). These studies show that UCNPs can be used as probes for targeted imaging of tumor cells, providing a powerful new tool for the diagnosis and treatment of tumors.

17.1.3 Multi-modality Imaging with UCNPs

Widely used clinical imaging techniques currently include magnetic resonance imaging (MRI), computed tomography (CT), positron emission tomography (PET), and single photon emission computed tomography (SPECT). These mono-modal imaging techniques reveal only a single aspect of the environment in the body. To obtain more relevant clinical information, multi-modal imaging technology has emerged, which combines several different imaging modes to overcome the limitations of individual methods (Fig. 17.2). In recent years, UCNPs-based multi-modal imaging technology has been rapidly developed and has received significant attention in the field of biomedical imaging. For example, MRI has high 3D spatial resolution, but

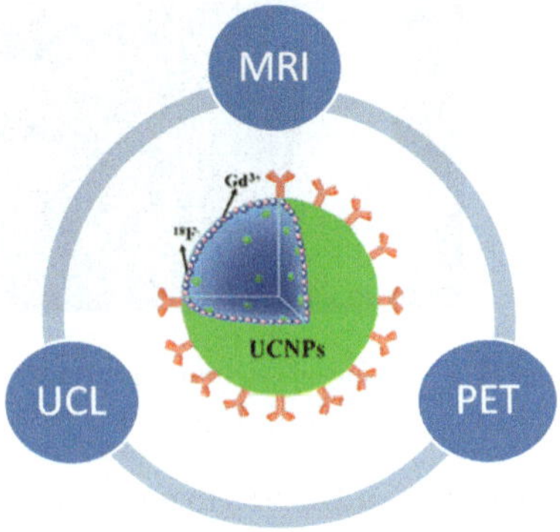

Fig. 17.2 UCNPs-based multi-modal imaging technology

has low imaging sensitivity; upconversion luminescence imaging (UCL) has high imaging sensitivity, but its spatial resolution is poor. A dual-mode imaging probe could provide complimentary advantages through combination of UCL with MRI, delivering enhanced biological imaging with high sensitivity and spatial resolution (Li et al. 2009; Xia et al. 2011).

Gadolinium ions (Gd^{3+}) are widely used as a contrast agent for MRI imaging. Hyeon (Park et al. 2009) doped Gd^{3+} into the matrix of UCNPs, successfully constructing a new dual-mode imaging probe ($NaGdF_4$:Yb/Er). The probe can be used for both UCL and MRI imaging of breast cancer cells (SK-BR3). It was also reported that UCL imaging can be combined with SPECT imaging (Liu et al. 2011; Sun et al. 2011). SPECT often uses the radioisotope ^{18}F as an imaging marker in clinical diagnosis and the composition of UCNPs includes the element of F. Therefore, a UCL/SPECT dual-mode imaging probe can be obtained by replacing F with ^{18}F when synthesizing UCNPs. By using this probe, high-quality in vivo imaging with high sensitivity, high spatial resolution, and good imaging penetration depth can be obtained in mice.

In addition, studies on multi-modal imaging, such as UCL/PET/MRI and UCL/CT/MRI, have been reported and these techniques are receiving increasing attention (Liu et al. 2013; Xia et al. 2012; Zhou et al. 2011). Combining multiple imaging technologies enables ultra-sensitive and multilayered imaging in systems ranging from cells to living organisms, which could not only improve the imaging accuracy, but also the diagnostic efficiency.

17.2 Application of UCNPs in Disease Treatment

17.2.1 Nanoscale Carriers for Drugs and Genes

The surface of UCNPs can be coated with mesoporous structures, allowing them to be used as carriers for small molecule drugs and genes for use in disease treatment.

UCNPs with mesoporous structure can be combined with drugs using electrostatic interaction (Tian et al. 2011). However, this approach usually increases the size of the nanomaterials due to the extra layer of silica on the particle surface. Another method is to modify the surface of the UCNPs with particular functional molecules. These functional molecules have high binding affinity with drug molecules, facilitating the binding of UCNPs and drugs.

Liu (Wang et al. 2011a) modified surface of UCNPs with doxorubicin (DOX), an anti-cancer chemotherapy drug, and applied the conjugate to tumor therapy. Because the water solubility of DOX is significantly enhanced under acidic conditions, and the microenvironment of the tumor extracellular tissue, intracellular lysosomes, and endosomes are all acidic; the amount of released DOX increases significantly when UCNPs reach the region of tumor cells, and further induces cell death. On this basis, UCNPs-DOX complexes can also be modified with FA. Thus, they can achieve targeted drug release in FA receptor positive cancer cells, which is of great clinical value for tumor therapy.

In addition, UCNPs can be transport carriers of DNA or RNA genes. It was reported that photolysable plasmid DNA/siRNA genes were linked to UCNPs with mesoporous structure (Jayakumar et al. 2012). And it was found that, it was possible to control the DNA/siRNA gene expression since the intensity of the excitation light can affect the gene expression in living cells.

17.2.2 Photodynamic Therapy

Photodynamic therapy (PDT) is an emerging new method for cancer treatment. Its principle is to transport chemical agents with light-activated properties (photosensitizers) around tumor cells (Fig. 17.3). Under light excitation, the photosensitizers can help to generate reactive oxygen species (ROS), which can induce the death of tumor cells (Wang et al. 2011b). The excitation wavelength of most photosensitizers is usually in the visible-near-infrared region, which has limited tissue penetrating ability and affects the treatment to a certain degree. The use of UCNPs as a photo-

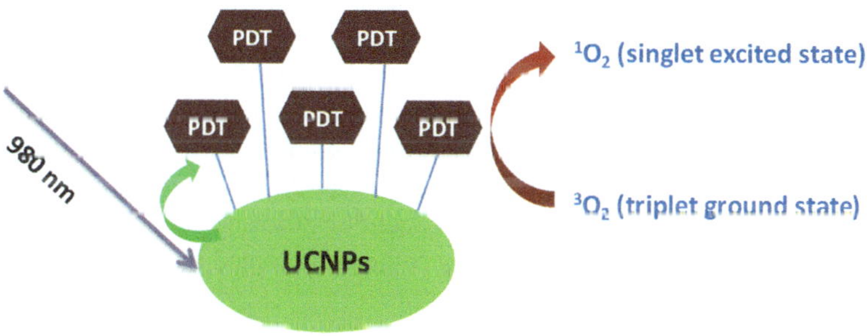

Fig. 17.3 Schematic description of UCNP-based PDT therapy

sensitizer carrier can effectively solve this problem. The excitation light (980 nm) of UCNPs has a deeper penetration depth, and the emitted light can directly expose the photosensitizer cargo. As a result, UCNPs have a better therapeutic effect on deep tumor cells. Moreover, UCNPs can be used for targeted PDT treatment by conjugating functional molecules (e.g. FA) on the surface.

By linking the photosensitizer (ZnPc) to UCNPs with mesoporous structure, Zhang successfully constructed a UCNP-based PDT nanoprobe (UCNPs-ZnPc) (Chatterjee and Yong 2008; Lim et al. 2012). The absorption peak of ZnPc (~670 nm) is in accordance with the red emission peak of UCNP materials. Therefore, under 980 nm excitation, ZnPc can absorb the red light emitted by UCNPs, and help to stimulate the generation of ROS to kill cancer cells. Their experiments showed that UCNPs-ZnPc could effectively reduce the activity of bladder cancer cells (MB49) in rat after 5 min under 980 nm laser excitation.

Moreover, it was reported that two or more different photosensitizers could be combined with UCNPs, and both could be stimulated by the light emitted by UCNPs. This approach could significantly improve the therapeutic effect on tumors (Idris et al. 2012). Other reported photosensitizers, such as Ce6 (Wang et al. 2011b), TPP (Shan et al. 2011), and TCPP (Wang et al. 2011a), can all be conjugated with UCNPs for PDT therapy.

17.2.3 Photothermal Therapy

UCNPs can also be applied in another emerging tumor therapy, photothermal therapy (PTT), which kills tumor cells by releasing heat. PTT therapy can affect tumor cells more selectively than PDT because the excitation light of the PTT agent (such as gold and silver particles) can irradiate tumors in the targeted region (Barreto et al. 2011). However, common endothermic materials have poor imaging quality in vivo, therefore, it is difficult to observe the treatment effect in real time. This problem could be resolved by using UCNPs as carriers of Au or Ag particles. Dong reported the preparation of core-shell UCNPs ($NaYF_4$:Yb, Er@Ag). The nanomaterial could be used for UCL imaging and PTT therapy with only a single 980 nm excitation laser (Dong et al. 2011). In the experiment, the material was incubated with human hepatoma cells (HepG2) in vitro. After 20 min of irradiation with the excitation light, the survival rate of tumor cells decreased from 65.1 to 4.6%, indicating a promising PTT treatment effect. In another report, Liu prepared multifunctional UCNP materials ($NaYF_4$:Yb/Er@Fe_3O_4@Au), which could be used for UCL/MRI dual-modal imaging and PTT therapy (Cheng et al. 2012). The animal experiment showed that tumor cells in a mouse model could be efficiently inhibited under near-infrared irradiation, and effective imaging could be obtained in vivo.

17.3 Biocompatibility and Toxicity of UCNPs

Currently, few nanomedicines are used in clinical practice, despite them being extensively studied for medical diagnosis and treatment. One of the reasons for this is the uncertain biocompatibility and cytotoxicity of nanomaterials. It is therefore important to systematically evaluate the toxicological characteristics and metabolic behavior of UCNPs in biological systems.

Many studies have shown that modified UCNPs materials have no apparent toxicity at the cellular level (Doane and Burda 2012; Yan et al. 2013). In one report, Bae and colleagues incubated a certain concentration of UCNPs with HeLa cells for a controlled period of time, and then used fluorescence imaging to study the spatial and temporal distribution of UCNPs in the HeLa cells. They found that UCNPs can enter cells via endocytosis and gather around the nucleus via the transportation of microtubule-associated motor proteins. Most of the UCNPs are eventually released from the cells through another type of motor protein. No significant cytotoxic effect on HeLa cells was observed (Bae et al. 2012). The composition of the main matrix ($NaYF_4$) of UCNP material is essentially non-toxic or low toxicity. However, it should be noted that the Gd^{3+} used in the UCL/MRI dual-mode imaging probe is toxic. To limit the toxicity, Gd^{3+} must be coupled with UCNPs in the form of chelates or coated forms.

In 2006, Li and coworkers first reported the in vivo toxicity of UCNPs against nematodes. They found that UCNPs could be significantly toxic to nematodes when the injection concentration was higher than 10 mg/ml (Lim et al. 2006). In another study, Zhang and coworkers injected silica-coated UCNPs (10 mg/kg) into mice via the tail vein (Abdul Jalil and Zhang 2008). After one day, the content of UCNPs in tissues rapidly decreased, while the content in the spleen was the highest. After 7 days, it was found that only a small amount of UCNP material remained in mice and no significant in vivo toxicity was observed during the experiment.

To further understand the chronic toxicity of UCNPs and their metabolism in vivo, Li and coworkers injected polyacrylic acid-modified UCNPs (PAA-UCNPs) into mice via intravenous injection, and observed their toxicity for 115 days. During the experiment, no obvious weight loss or other abnormalities were observed in the mice. The imaging displayed that UCNPs were mainly concentrated in the liver and spleen of the mice, and most of them were slowly excreted through metabolic pathways. In addition, blood biochemical analysis and histopathological analysis of tissue sections, showed that PAA-UCNP treatment did not result in significant toxicity in mice. These results demonstrated that UCNP material has good biocompatibility and low cytotoxicity for in vivo imaging and treatment. However, there are still some problems with the toxicity of UCNPs. For example, the interaction of UCNPs with stem cells and immune cells, and the immunological effects on the body remain unclear. Further toxicological studies on UCNP nanomaterials require more systematic and in-depth research.

17.4 Conclusion

UCNPs nanomaterials have a wide range of applications in biomedicine because of their unique luminescence properties. Over the past few years, significant developments in the synthesis, structure optimization, and surface modification of UCNPs have been achieved. However, there are still some limitations that must be researched and explored.

First, although considerable achievements have been made in the preparation of UCNPs, obtaining UCNPs with high luminescence efficiency is still a crucial challenge that is important for the further improvement of UCNPs in biomedical applications. Second, the development of UCNPs as drug delivery carriers is still in the primary stages and there remain many challenges in establishing an effective, reliable, and smart nanoparticle-based drug delivery system. For example, how to release drugs accurately and controllably. Finally, there remains a need for more systematic and comprehensive research into the toxicity of UCNPs. As a new type of biomedical material, UCNP nanomaterials will face complex challenges in both basic theory and practical application, which will require researchers in various disciplines to work closely together. Overcoming these challenges will allow UCNP nanomaterials to play a more significant role in medical research and application.

References

Abdul Jalil R, Zhang Y. Biocompatibility of silica coated NaYF(4) upconversion fluorescent nanocrystals. Biomaterials. 2008;29(30):4122–4128.

Bae YM, Park YI, Nam SH, Kim JH, Lee K, Kim HM, Yoo B, Choi JS, Lee KT, Hyeon T, Suh YD. Endocytosis, intracellular transport, and exocytosis of lanthanide-doped upconverting nanoparticles in single living cells. Biomaterials. 2012;33(35):9080–9086.

Barreto JA, O'Malley W, Kubeil M, Graham B, Stephan H, Spiccia L. Nanomaterials: applications in cancer imaging and therapy. Adv Mater. 2011;23(12):H18–H40.

Chatterjee DK, Yong Z. Upconverting nanoparticles as nanotransducers for photodynamic therapy in cancer cells. Nanomedicine (Lond). 2008;3(1):73–82.

Cheng L, Yang K, Li Y, Zeng X, Shao M, Lee ST, Liu Z. Multifunctional nanoparticles for upconversion luminescence/MR multimodal imaging and magnetically targeted photothermal therapy. Biomaterials. 2012;33(7):2215–2222.

Doane TL, Burda C. The unique role of nanoparticles in nanomedicine: imaging, drug delivery and therapy. Chem Soc Rev. 2012;41(7):2885–2911.

Dong B, Xu S, Sun J, Bi S, Li D, Bai X, Wang Y, Wang L, Song H. Multifunctional NaYF4: Yb3+, Er3+@Ag core/shell nanocomposites: integration of upconversion imaging and photothermal therapy. J Mater Chem. 2011;21(17):6193.

Idris NM, Gnanasammandhan MK, Zhang J, Ho PC, Mahendran R, Zhang Y. In vivo photodynamic therapy using upconversion nanoparticles as remote-controlled nanotransducers. Nat Med. 2012;18(10):1580–1585.

Jayakumar MK, Idris NM, Zhang Y. Remote activation of biomolecules in deep tissues using near-infrared-to-UV upconversion nanotransducers. Proc Natl Acad Sci USA. 2012;109(22):8483–8488.

Li Z, Zhang Y, Shuter B, Muhammad Idris N. Hybrid lanthanide nanoparticles with paramagnetic shell coated on upconversion fluorescent nanocrystals. Langmuir. 2009;25(20):12015–12018.

Lim SF, Riehn R, Ryu WS, Khanarian N, Tung CK, Tank D, Austin RH. In vivo and scanning electron microscopy imaging of up-converting nanophosphors in Caenorhabditis elegans. Nano Lett. 2006;6(2):169–174.

Lim ME, Lee YL, Zhang Y, Chu JJ. Photodynamic inactivation of viruses using upconversion nanoparticles. Biomaterials. 2012;33(6):1912–1920.

Liu Q, Sun Y, Li C, Zhou J, Yang T, Zhang X, Yi T, Wu D, Li F. 18F-Labeled magnetic-upconversion nanophosphors via rare-earth cation-assisted ligand assembly. ACS Nano. 2011;5(4):3146–3157.

Liu Z, Pu F, Huang S, Yuan Q, Ren J, Qu X. Long-circulating Gd(2)O(3):Yb(3+), Er(3+) up-conversion nanoprobes as high-performance contrast agents for multi-modality imaging. Biomaterials. 2013;34(6):1712–1721.

Nyk M, Kumar R, Ohulchanskyy TY, Bergey EJ, Prasad PN. High contrast in vitro and in vivo photoluminescence bioimaging using near infrared to near infrared up-conversion in Tm^{3+} and Yb^{3+} doped fluoride nanophosphors. Nano Lett. 2008;8(11):3834–3838.

Park YI, Kim JH, Lee KT, Jeon K-S, Na HB, Yu JH, Kim HM, Lee N, Choi SH, Baik S-I, Kim H, Park SP, Park B-J, Kim YW, Lee SH, Yoon S-Y, Song IC, Moon WK, Suh YD, Hyeon T. Nonblinking and nonbleaching upconverting nanoparticles as an optical imaging nanoprobe and T1 magnetic resonance imaging contrast agent. Adv Mater. 2009;21(44):4467–4471.

Shan J, Budijono SJ, Hu G, Yao N, Kang Y, Ju Y, Prud'homme RK. Pegylated composite nanoparticles containing upconverting phosphors and meso-tetraphenyl porphine (TPP) for photodynamic therapy. Adv Func Mater. 2011;21(13):2488–2495.

Sun Y, Yu M, Liang S, Zhang Y, Li C, Mou T, Yang W, Zhang X, Li B, Huang C, Li F. Fluorine-18 labeled rare-earth nanoparticles for positron emission tomography (PET) imaging of sentinel lymph node. Biomaterials. 2011;32(11):2999–3007.

Tian G, Gu Z, Liu X, Zhou L, Yin W, Yan L, Jin S, Ren W, Xing G, Li S, Zhao Y. Facile fabrication of rare-earth-doped Gd_2O_3 Hollow spheres with upconversion luminescence, magnetic resonance, and drug delivery properties. J Phys Chem C. 2011;115(48):23790–23796.

Wang C, Cheng L, Liu Z. Drug delivery with upconversion nanoparticles for multi-functional targeted cancer cell imaging and therapy. Biomaterials. 2011a;32(4):1110–1120.

Wang C, Tao H, Cheng L, Liu Z. Near-infrared light induced in vivo photodynamic therapy of cancer based on upconversion nanoparticles. Biomaterials. 2011b;32(26):6145–6154.

Xia A, Gao Y, Zhou J, Li C, Yang T, Wu D, Wu L, Li F. Core-shell NaYF4:Yb3+, Tm3+@FexOy nanocrystals for dual-modality T2-enhanced magnetic resonance and NIR-to-NIR upconversion luminescent imaging of small-animal lymphatic node. Biomaterials. 2011;32(29):7200–7208.

Xia A, Chen M, Gao Y, Wu D, Feng W, Li F. Gd3+ complex-modified NaLuF4-based upconversion nanophosphors for trimodality imaging of NIR-to-NIR upconversion luminescence, X-Ray computed tomography and magnetic resonance. Biomaterials. 2012;33(21):5394–5405.

Xiong LQ, Chen ZG, Yu MX, Li FY, Liu C, Huang CH. Synthesis, characterization, and in vivo targeted imaging of amine-functionalized rare-earth up-converting nanophosphors. Biomaterials. 2009a;30(29):5592–5600.

Xiong L, Chen Z, Tian Q, Cao T, Xu C, Li F. High contrast upconversion luminescence targeted imaging in vivo using peptide-labeled nanophosphors. Anal Chem. 2009b;81(21):8687–8694.

Yan L, Chang Y-N, Zhao L, Gu Z, Liu X, Tian G, Zhou L, Ren W, Jin S, Yin W, Chang H, Xing G, Gao X, Zhao Y. The use of polyethylenimine-modified graphene oxide as a nanocarrier for transferring hydrophobic nanocrystals into water to produce water-dispersible hybrids for use in drug delivery. Carbon. 2013;57:120–129.

Yang T, Sun Y, Liu Q, Feng W, Yang P, Li F. Cubic sub-20 nm NaLuF(4)-based upconversion nanophosphors for high-contrast bioimaging in different animal species. Biomaterials. 2012;33(14):3733–3742.

Yu M, Li F, Chen Z, Hu H, Zhan C, Yang H, Huang C. Laser scanning up-conversion luminescence microscopy for imaging cells labeled with rare-earth nanophosphors. Anal Chem. 2009;81(3):930–935.

Yu XF, Sun Z, Li M, Xiang Y, Wang QQ, Tang F, Wu Y, Cao Z, Li W. Neurotoxin-conjugated upconversion nanoprobes for direct visualization of tumors under near-infrared irradiation. Biomaterials. 2010;31(33):8724–8731.

Zhou J, Yu M, Sun Y, Zhang X, Zhu X, Wu Z, Wu D, Li F. Fluorine-18-labeled Gd3+/Yb3+/Er3+ co-doped NaYF4 nanophosphors for multimodality PET/MR/UCL imaging. Biomaterials. 2011;32(4):1148–1156.

The manufacturer's authorised representative in the EU is Springer Nature Customer Service Centre GmbH, Europaplatz 3, 69115 Heidelberg, Germany. If you have any concerns regarding our products, please contact ProductSafety@springernature.com

Printed and bound by CPI Group (UK) Ltd, Croydon, CR0 4YY
23/07/2026
02175005-0004